FIVE HUNDRED POINTS OF
GOOD HUSBANDRY

Five Hundred Points of Good Husbandry

THOMAS TUSSER

With an introduction by
GEOFFREY GRIGSON

Oxford New York
OXFORD UNIVERSITY PRESS
1984

Oxford University Press, Walton Street, Oxford OX2 6DP
London Glasgow New York Toronto
Delhi Bombay Calcutta Madras Karachi
Kuala Lumpur Singapore Hong Kong Tokyo
Nairobi Dar es Salaam Cape Town
Melbourne Auckland
and associated companies
Beirut Berlin Ibadan Mexico City Nicosia

Oxford is a trade mark of Oxford University Press

Introduction © Geoffrey Grigson 1984
Index and revised glossary © Oxford University Press 1984

For publication history see Publisher's Note p. xxi
First published as an Oxford University Press paperback 1984

All rights reserved. No part of this publication may be reproduced,
stored in a retrieval system, or transmitted, in any form or by any means,
electronic, mechanical, photocopying, recording, or otherwise, without
the prior permission of Oxford University Press

This book is sold subject to the condition that it shall not, by way
of trade or otherwise, be lent, re-sold, hired or otherwise circulated
without the publisher's prior consent in any form of binding or cover
other than that in which it is published and without a similar condition
including this condition being imposed on the subsequent purchaser

British Library Cataloguing in Publication Data
Tusser, Thomas
Five hundred points of good husbandry.—(Oxford paperbacks)
1. Agriculture—Early works to 1600
I. Title II. Grigson, Geoffrey
630'.9 S435
ISBN 0-19-286040-2

Printed in Great Britain by
Richard Clay (The Chaucer Press) Ltd.
Bungay, Suffolk

CONTENTS

Introduction	xi
Publisher's Note	xxi

FIVE HUNDRED POINTS OF GOOD HUSBANDRY

The authors epistle to the late Lord William Paget	1
To the Right Honourable the Lord Thomas Paget of Beaudesert	2
To the reader	7
An introduction to the booke of husbandrie	9
A preface to the buyer	10
The commodities of husbandrie	11
The description of husbandrie	12
The ladder to thrift	13
Good husbandlie lessons	15
The fermers dailie diet	23
A description of the properties of windes	25
Of the planets	26
Septembers abstract	27
Septembers husbandrie	30
Octobers abstract	38
Octobers husbandrie	41
Novembers abstract	47
Novembers husbandrie	49
Decembers abstract	53
Decembers husbandrie	55
A digression to hospitalitie	58
A description of time, and the yeare	59

A description of life and riches	60
A description of housekeeping	61
A description of the feast and birth of Christ	62
A description of apt time to spend	63
Against fantasticall scruplenes	64
Christmas husbandlie fare	65
A Christmas caroll	66
Januaries abstract	68
Of trees or fruites to be set or removed	72
Januaries husbandrie	73
Februaries abstract	80
Februaries husbandrie	83
Marches abstract	86
Seedes and herbes for the kitchen	88
Herbes and roots for sallets and sauce	89
Herbes and rootes to boile or to butter	89
Strowing herbes of all sortes	90
Herbes, branches, and flowers, for windowes and pots	90
Herbes to still in sommer	91
Necessarie herbes to growe in the garden for physick	91
Marches husbandrie	92
Aprils abstract	96
Aprils husbandrie	97
A lesson for dairie maid Cisley	100
Maies abstract	102
Maies husbandrie	104
Junes abstract	109
Junes husbandrie	110

A lesson where and when to plant good hopyard	113
Julies abstract	114
Julies husbandrie	115
Augusts abstract	117
Augusts husbandrie	121
Works after harvest	124
Corne harvest	128
A briefe conclusion	129
Mans age divided by prentiships, from birth to his grave	130
Another division of the nature of mans age	131
Comparing good husband with unthrift his brother	132
A comparison betweene champion countrie and severall	134
The description of an envious and naughtie neighbour	140
To light a candell before the Devill	141
A sonet against a slanderous tongue	144
A sonet upon the authors first seven yeares service	145
The authors dialogue betweene two bachelers	146
THE POINTS OF HUSWIFERIE	153
To the Right Honorable the Ladie Paget	154
To the reader	156
The preface to the booke of huswiferie	157
A description of huswife and huswiferie	158
Instructions to huswiferie	159
Morning workes	163
Breakefast doings	164
Huswifely admonitions	164
Brewing	167

Baking	167
Cookerie	168
Dairie	168
Scouring	169
Washing	169
Malting	170
Dinner matters	170
After noone workes	172
Evening workes	174
Supper matters	175
After supper matters	176
The ploughmans feasting daies	177
The good huswifelie physicke	179
The good motherlie nurserie	180
A comparison between good huswiferie and evill	181

*

For men a perfect warning how childe shall come by larning	182
The description of a womans age	184
The inholders posie	184
Certaine table lessons	185
Lessons for waiting servants	185
Husbandly posies for the hall	186
Posies for the parler	187
Posies for the gests chamber	187
Posies for thine owne bed chamber	188
A sonet to the Ladie Paget	189
Principall points of religion	190
The authors beliefe	191

Of the omnipotencie of God	196
Eleemosyna prodest homini in vita, in morte, & post mortem	197
Malus homo	198
Of two sorts of men	198
Diabolo cum resistitur, est ut formica	198
Eight of S. Barnards verses	199
Of the authors linked verses departing from Court to the Countrie	201
The authors life	202
Fortuna non est semper amica	212
Notes and illustrations	214
Appendix: Sir Walter Scott's introduction as reprinted in the 1931 Tregaskis edition	315
Glossary	321
Index	337

INTRODUCTION

THOMAS TUSSER (1524?–1580)

MORE than four centuries have gone by since Thomas Tusser Gent., an East Anglian leaseholder without an estate of his own, wrote a doggerel calendar of the farmer's year, first published for him in black letter, in 1557, by Richard Tottel, the famous 'stationer' who collected poems, great poems, some of them, and poems by aristocratic poets, into *Tottel's Miscellany*.

The edition of 1557 was called *A hundredth goode pointes of husbandrie*. He married, so for the benefit of wives – and husbands no doubt with wives to discipline – he added to later editions 'a hundrethe good pointes of Huswifry'. The hundred points of husbandry went up to five hundred in the edition of 1573, and as edition followed edition the advice Thomas Tusser had to give was variously augmented and corrected, drawing on his farming experience in Essex, Suffolk, and Norfolk and his views on life and religion.

It is Tusser's book which tells us most of what we know about him, most too of what we can guess or deduce about his character, thanks especially to 'The Author's Life', in forty stanzas of autobiography, which he added to the second edition of 1570.

The Tussers lived in the Essex parish of Rivenhall, north of the Chelmsford–Colchester highway, and about eight miles from the saltings of Maldon and the Blackwater. They were a family of no distinction, no particular fortune, so Thomas Tusser was naively pleased to mention – as if assuring readers that they could safely read his book – that short as their pedigree was, the Tussers were included by the College of Heralds in their Visitation of Essex. As the youngest son in a family of five – four boys and a girl – Thomas would have to make his own way, with whatever help his parents could find.

Luckily – or as he may have thought, unluckily – this boy could sing, and obviously he could sing very well, in a time which cared increasingly about church music and madrigals and song books. His father evidently saw that singing could procure for him both education and livelihood. My guess is that this young Tusser had perfect pitch and that someone, perhaps some cleric from Cambridge, told his father how rare this was and how much of a life insurance, or at least a youth insurance, it could be. My second guess is that Thomas Tusser, whom we incline to think of sentimentally and rather patronizingly as a portion of the quaint olde world, must be thought of as once a slim, attractive boy, much attached to his home, his family, and his 'village faier', who liked nothing so much as pottering about in the land of his infancy – a boy not excessively bright, who sang around the house or the fields for light-hearted pleasure, like the flaxen-headed cowboy in the folksong.

In an awkward stanza in that 'Life of Tusser' he seems to say that he had never been parted from his mother, but that his father insisted. With that voice he was to be a chorister. His father managed to place the boy miles away from home –

> But out I must, to song be thrust
> Say what I would, do what I could,
>   his mind was so –

in the collegiate chapel of the once great castle of Wallingford, in Berkshire, on the bank of the Thames. So began the agonies of a young songster whose one desire was to continue a country life in Essex.

> O painfull time, for everie crime,
> What toesed[1] ears, like baited beares!
> What bobbed lips, what jerks, what nips!
>   What toies[2]!

[1] *toesed*: tweaked   [2] *toies* (toys): tricks, jokes

> What robes, how bare! What colledge fare!
> What bread, how stale! What pennie Ale!
> When Wallingford, how wort thou abhord
> of sillie boies!

Tusser's release was to become, happily enough, a chorister at St Paul's, under the Master of the Children of St Paul's, John Redford, teacher and composer of distinction:

> – the like no where,
> For cunning such, and vertue much
> By whom some part of Musicke art
> so did I gaine.

From St Paul's Tusser was sent to Eton under the remarkable, scholarly, severe headmaster Nicholas Udall. This was a descent again to hell – at least to lifelong recollection of fifty-three strokes administered at one go 'for fault but small, or none at all.' 'That was the mercie', he wrote, extended by Udall to a poor boy. My third guess is that the fault but small might have been boneheadedness in learning his Latin phrases. Udall would have preferred intelligence to charm, if that was really Tusser's strong suit.

Tusser was now about nineteen, and fifty-three strokes or no, his fortune was changing through what may have been some long-standing family connection with a powerful cultured nobleman, Lord Paget, Secretary of State and Knight of the Garter. Was it Paget who had procured him a place at St Paul's, where Paget had been schooled? At Cambridge before long, was it on Paget's advice that Tusser gave up his place at King's College and transferred to Trinity Hall, which had been Paget's college? Paget was then High Steward of the University.

After Eton and a spell in London, Tusser at last found Trinity Hall little short of heaven:

> From London hence, to Cambridge thence
> With thanks to thee, O Trinitee,
> That to thy hall, so passing all
>   I got at last:
>
> There joy I felt, there trim I dwelt,
> There heaven from hell, I shifted well,
> With learned men, a number then,
>   the time I past.

Anyhow, when he left Cambridge, Lord Paget made the young Tusser his musician at court and presumably in his household at Beaudesert in Staffordshire, near Lichfield. The great mansion of Beaudesert has been destroyed, leaving only a portion of the Great Hall, which may have heard Tusser's voice. Serving a patron he had reason to love, Tusser sang for his victuals for ten years, all the time yearning, we may suppose, for his lost Essex home and the Essex farm land.

> My serving you, (thus understand,)
> And God his helpe, and yours withall,
> Did cause good lucke to take mine hand
> Erecting one most like to fall.

He took to a new music, he says, 'My Musicke since hath bene the plough'; and he confessed years later in a second dedication of his much enlarged book, this time to Lord Paget's son, the second lord, that he ought to have taken the first Lord Paget's advice not to leave court and music for the land, which had led him into nothing but misfortune.

He seems to have amassed enough money for this move to independence, marrying, and setting up first of all, not in Essex, but in Suffolk, just over the border in a hall house in the hamlet of Cattiwade. Cattiwade is in the parish of Brantham, and near Brantham is Willy Lot's mill, Flatford Mill, painted so famously two and a half centuries later by John Constable. So we may go into an art gallery and visualize Thomas

Tusser's ox teams dragging plough and harrow and wagon across a Constable landscape of green serenity, much more delectable than the bare and flattish acres of his native Rivenhall, twenty miles away.

Troubles soon began. Lover of farming or no, this musician returned to his childhood seems to have lacked after all the aptitude and energy and business sense required to make a go of farming:

> There was I faine my selfe to traine
> To learne too long the fermers song
> For hope of pelfe, like worldly elfe
>   to moile and toil.

He suffered 'loss and paine, to little gaine'. Worse followed. His wife could not stand the dampness of the Stour valley, and sickened. Tusser, who had already been attacked by malaria, gave up his farm on the edge of the saltings and moved a few miles inland to Ipswich, which he described as

> A towne of price, like paradice,
> For quiet then and honest men.

But his wife did not improve, and he soon found himself a widower. Before long he married a girl from Norfolk, who gave him the cares and costs of a young wife and a family of four children. It seems to have been this Norfolk connection which took him to West Dereham, where a new landlord settled him comfortably in a home in what had been the precinct of the great Premonstratensian abbey of West Dereham, which had been dissolved only some twenty years before. He was happy in this

> place for wood, that trimlie stood,
> With flesh and fish, as heart would wish,

but quarrelling between the landlords unsettled him:

> but when I spide
> That Lord with Lord could not accord,
> But now pound he, and now pound we,
> Then left I all, bicause such brall,
>   I list not bide,

and he tried a new life once more; this time, it seems, under the protection of Sir Richard Southwell, one of the wealthiest grandees of England and East Anglia, a courtier whose grandson was Robert Southwell, the poet who wrote 'The Burning Babe', and whose features, as a young man, clean-shaven and thoughtfully aloof, we know from Holbein's drawing at Windsor Castle and Holbein's portrait of him in the Uffizi (Holbein had also painted Lord Paget).

Tusser was out of luck again, Lord Paget his first patron, so loved and loving, died in 1563, and now Southwell died in 1564:

> O death thou so, why didst thou so
> Ungently treat that Jewell great,
> Which opte his doore to rich and poore,
>   so bounteously?

Away he went, this time to Norwich, where he was helped by yet another patron, John Salisbury, Dean of the cathedral –

> Thou gentle deane, mine only mean,
>   there then to live –

who presumably employed him on the cathedral music and choir. He was ill in Norwich, from a stoppage of his bladder so severe, he records in the margin of his verse account of himself, that 'in 138 houres I never made drop of water'.

Yet he was not finished with farming. He left Norwich, and back in Essex, in his own county, took on the working of the parson's glebe lands at Fairstead. This was going home, or it should have been, since Fairstead is only a few miles from the Rivenhall of Tusser's nativity. But no. He was too tired, too unwell, 'with sickness worne, as one forlorn'. He ran for London. The plague swept through London and the Tussers were given refuge in Cambridge, by his old college Trinity Hall. Music became his plough once more. The end of the matter was a return to London when the plague slackened, and there he died on 8 May 1580, in prison for debt, a failure, not far from sixty, which was a fair age for Tusser's times – a failure, a floater, according to the brief account of him by Thomas Fuller in his *Worthies of Britain*, who had been grazier (cattle farmer, that is) as well as husbandman, a renter who 'impoverished himself, and never enriched his landlord', a trader who had 'traded at large in oxen, sheep, dairies (milch-cattle), grain of all kinds. Whether he bought or sold, he lost.'

A failure who had been cherished, to little effect, by so many in high places, who evidently loved him and his singing, but were powerless to reshape his character.

Tusser wrote the first version of his book – or rather, as it became, his miscellany – when he was still farming at Cattiwade, in that Constable country. He was in his early thirties then, failure or misfortune had yet to become the fixed feature of his life.

Why was he writing a book? Was he saying to Lord Paget, who had advised him against farming and against giving up his music and his life at court (which Tusser was to condemn as an actually and morally dangerous milieu of 'Cards and Dice, with Venus vice And peevish pride'), 'You see I do know what I am doing'? No one has ever pretended that this farmer-musician or musician-farmer could write other than abominably. Was he in fact one of those men of music whose culture doesn't extend a millimetre beyond their single art?

He was never short of a patron, rich and cultured. One after another – to renew guessing, after all – was he being humoured by these patrons because he sang so well and had given them such pleasure, and was so likeable? In England certainly he was something of a Renaissance pioneer, he was adapting and extending a classical tradition. Before that first version of his *Pointes of Husbandrie* appeared, there had been nothing in the agricultural line, in English, but Fitzherbert's *Book of Husbandrie*, of which seven editions had been printed before that 1557 edition of the *Pointes*. Had Nicholas Udall flogged Virgil's *Georgics* into him at Eton, or tried to?

Much happened to him in the thirteen years between the *hundredth goode pointes of husbandrie* of 1557 and the second edition of 1570 which he enlarged with his 'hundreth good poynts of huswifery'. The long and short of it was that sad progress

> To carke and care, and ever bare,
> With losse and paine, to little gaine,

accentuated by the death of his Lord Paget in 1563, when he was thirty-nine, and by his removal to Ipswich and the death of his first wife. Cark and care not lessening, gaine not improving, it looks as if he fell back on his poem – his miscellany – enlarging it, adding recollection, and comforting himself in sententious and pious and slightly less awkward verse –

> This is my stedfast Creede, my faith, and all my trust,
> That in the heavens there is a God, most mightie, milde
>   and just

– all of which he swept into his now *Five hundreth points of good husbandry united to as many of good huswiferie* of 1573, with extras in one edition and another: a Tudor best-seller, it proved, of which there followed no fewer than six editions in his last seven years.

Reprintings and more reprintings went on being called for after Tusser's death well into the new century. Why? Certainly the expression of his didactic verse was odd by the improving standards of Elizabethan prose and the fresh euphony and grace of Elizabethan verse, but here was unquestionably helpful advice on the gamut of farming; which, after all, was the basis of life, wealth, and power. Here was a practical and moralistic compendium for an English land-based society turning its back on the Middle Ages. A host of new landlords and cultivators were enjoying and exploiting lands retrieved from monastic ownership and from a neglectful backward-looking tenantry.

Yet more is needed to explain the lasting popularity of Tusser's cheerful, virtuous, once useful and novel admonition, and the fact that like the late Tudors we too can read Tusser – awkwardness or no; broken rhythms, bad grammar, abominable syntax, naivety, or no – with unexpected pleasure. In economic history, social history, agricultural history of course the *Five Hundreth Points* is, and is going to remain, a document of value. But in discovering from this 'Stone of Sisyphus', as Thomas Fuller called him, who rolled around and gathered no moss for himself, we can, if pre-industrial country life interests us, savour the pure enjoyment of seeing how farms looked in these Tudor years, how farming and farmers' wives proceeded. Here is the farmer sitting to his food, according to season: to his salt herring and haberdine, or special grade of salt cod, at Lent; to his Martinmas beef – beef which had been chimney-smoked at the previous Martinmas – and his veal and bacon at Easter, to his mackerel next; then at mid-summer to his fresh 'grassebeefe' and peas; at Michaelmas to fresh herrings and 'fatted Crones' – which were old fattened ewes no longer good for lambing; and at All Saints from new killed pigs to his pork and souse – pig's trotters and pig's ears – and the sprats and the spurlings or smelts, which, like the Yarmouth herrings, were such a sea-crop of the East Anglian fishermen.

Allow our eyes a little extra vision and Tusser takes us to ploughing and fallowing, to weeding the growing crops in May and June, ridding them as much as the farmer can from poppies and corncockle and boddles (corn marigolds), from fitchis (vetches), bracken, and that terrible mayweed (*Anthemis cotula*), which blistered the arms and naked chests and shoulders and backs of the reapers. As well as the sowing and harvesting of wheat and barley, Tusser allows us to see how hemp was grown and cultivated and prepared. We are taken to sheep-washing and shearing, to gelding of rams and bulls, to mole-catching, to cutting out planks in the sawpit, to inspection of the thatch, removing the moss which grew on it through the year, and smoothing the thatch into place. Wives are sent to the woods for strawberry plants,

> Wife, into thy garden, and set me a plot,
>   with strawbery rootes, of the best to be got,

necessary vegetables are listed, and the best plants for strewing as carpet-cover over the cold floors. Again and again there is instructive mention of how to grow saffron, that East Anglian speciality of others than the saffron farmers or 'crokers' around Saffron Walden (and around Stratton in Cornwall).

Rightly approached and read, this peculiar miscellany becomes a small encyclopaedia of old farming, a special dictionary of intriguing words (words, by the way, which often occur in place-names or at least in field names) — all of that over and above the enigmas presented by Tusser himself and his music and the pleasures and disappointments of his half-metropolitan, half-rural career. He is a generous writer, this man who loved husbandry, forcing himself to wrestle with the intractability of language —

> Now looke up to Godward, let tong never cease
>   in thanking of him, for his mightie encrease:
> Accept my good will, for a proof go and trie:
>   the better thou thrivest, the gladder am I.

PUBLISHER'S NOTE

THE text is that of the edition of 1580, the last edition issued in Tusser's lifetime, edited by Henry Denham.

An edition of this text was published in 1878 by the (now defunct) English Dialect Society with a glossary and notes, and this edition was reissued in 1931 by James Tregaskis and Son with an introduction by Sir Walter Scott and additional notes from earlier editions.

The Tregaskis edition has been reproduced photographically for this Oxford University Press paperback edition with a new introduction by Geoffrey Grigson, who has also revised the glossary. An index has also been added.

FIVE HUNDRED POINTS
of
GOOD HUSBANDRY

THE AUTHORS EPISTLE
to the late Lord William Paget,
wherein he doth discourse of his own bringing up,
and of the goodness of the said Lord his master unto him,
and of the occasion of this his booke
thus set forth of his owne long practise.

Chapter 1

T   *Time trieth the troth, in everie thing,*
H   *Herewith let men content their minde,*
O   *Of works, which best may profit bring,*
M   *Most rash to judge, most often blinde.*
A   *As therefore troth in time shall crave,*
S   *So let this booke just favor have.*

T   *Take you my Lord and Master than,*
U   *Unlesse mischance mischanceth me,*
S   *Such homelie gift, of me your man,*
S   *Since more in Court I may not be,*
A   *And let your praise, wonne heretofore,*
R   *Remaine abrode for evermore.*

M   *My serving you, (thus understand,)*
A   *And God his helpe, and yours withall,*
D   *Did cause good lucke to take mine hand,*
E   *Erecting one most like to fall.*

M   *My serving you, I know it was,*
E   *Enforced this to come to pas.*

Since being once at Cambridge taught,
Of Court ten yeeres I made assaie,
No Musicke then was left unsaught,
Such care I had to serve that waie.
When joie gan slake, then made I change,
Expulsed mirth, for Musicke strange.

My Musicke since hath bene the plough,
Entangled with some care among,
The gaine not great, the paine ynough,
Hath made me sing another song.
Which song, if well I may avow,
I crave it judged be by yow.

     Your servant Thomas Tusser.

TO THE RIGHT HONORABLE
and my speciall good Lord and Master,
THE LORD THOMAS PAGET OF BEAUDESERT
Son and heire to his late
father deceased.

Chapter 2

MY Lord, your father looved me,
   and you my Lord have prooved me,
and both your loves have mooved me,
   to write as here is donne:
Since God hath hence your father,
   such flowers as I gather,
I dedicate now rather,
   to you my Lord his sonne.

Your father was my founder,
till death became his wounder,
no subject ever sounder,
   whome Prince advancement gave:
As God did here defend him,
and honour here did send him,
so will I here commend him,
   as long as life I have.

His neighbours then did blisse him,
his servants now doe misse him,
the poore would gladlie kisse him,
   alive againe to be:
But God hath wrought his pleasure,
and blest him, out of measure,
with heaven and earthlie treasure,
   so good a God is he.

Ceres the Goddesse of husbandrie.

His counsell had I used,
and *Ceres* art refused,
I neede not thus have mused,
   nor droope as now I do:
But I must plaie the farmer,
and yet no whit the warmer,
although I had his armer,
   and other comfort to.

Æsops fable

The Foxe doth make me minde him,
whose glorie so did blinde him,
till taile cut off behinde him,
   no fare could him content:
Even so must I be prooving,
such glorie I had in looving,
of things to plough behooving,
   that makes me now repent.

Salust.

Loiterers I kept so meanie,
both Philip, Hob, and Cheanie,
that, that waie nothing geanie,
   was thought to make me thrive:
Like *Iugurth*, Prince of *Numid*,
my gold awaie consumid,
with losses so perfumid,
   was never none alive.

Great fines so neere did pare me,
great rent so much did skare me,
great charge so long did dare me,
   that made me at length crie creake:
Much more of all such fleeces,
as oft I lost by peeces,
among such wilie geeces
   I list no longer speake.

Though countrie health long staid me,
yet lesse expiring fraid me,
and (*ictus sapit*) praid me
    to seeke more steadie staie:
New lessons then I noted,
and some of them I coted,
least some should think I doted,
    by bringing naught awaie.

Though *Pallas* hath denide me,   *Pallas,*
hir learned pen to guide me,   *Goddesse of*
for that she dailie spide me,   *wisdome and*
    with countrie how I stood: *cunning.*
Yet *Ceres* so did bold me,
with hir good lessons told me,
that rudenes cannot hold me,
    from dooing countrie good.

By practise and ill speeding,
these lessons had their breeding,
and not by hearesaie, or reeding,
    as some abrode have blowne:
Who will not thus beleeve me,
so much the more they greeve me,
because they grudge to geeve me,
    that is of right mine owne.

At first for want of teaching,
at first for trifles breaching,
at first for over reaching,
    and lacke of taking hede,
was cause that toile so tost me,
that practise so much cost me,
that rashnes so much lost me,
    or hindred as it did.

Yet will I not despaier
thorough God's good gift so faier
through friendship, gold, and praier,
   in countrie againe to dwell:
Where rent so shall not paine me,
but paines shall helpe to gaine me,
and gaines shall helpe maintaine me,
New lessons mo to tell.

For citie seemes a wringer,
the penie for to finger,
from such as there doe linger,
   or for their pleasure lie:
Though countrie be more painfull,
and not so greedie gainfull,
yet is it not so vainfull,
   in following fansies eie.

I have no labour wanted
to prune this tree thus planted,
whose fruite to none is scanted,
   in house or yet in feeld:
Which fruite, the more ye taste of,
the more to eate, ye haste of,
the lesse this fruite ye waste of,
   such fruite this tree doth yeeld.

My tree or booke thus framed,
with title alreadie named,
I trust goes forth unblamed,
   in your good Lordships name:
As my good Lord I take you,
and never will forsake you,
so now I crave to make you
   defender of the same.

      Your servant Thomas Tusser.

TO THE READER

Chapter 3

I HAVE been praid
to shew mine aid,
in taking paine,
not for the gaine,
but for good will,
to shew such skill
  as shew I could:
That husbandrie
with huswiferie
as cock and hen,
to countrie men,
all strangenes gone,
might joine in one,
as lovers should.

I trust both this
performed is,
and how that here
it shall appere,
with judgement right,
to thy delight,
  is brought to passe:
That such as wive,
and faine would thrive,
be plainly taught
how good from naught
may trim be tride,
and lively spide,
  as in a glasse.

What should I win,
by writing in
my losses past,
that ran as fast
as running streame,
from reame to reame
  that flowes so swift?
For that I could
not get for gould,
to teach me how,
as this doth yow,
through daily gaine,
the waie so plaine
  to come by thrift.

What is a grote
or twaine to note,
once in the life
for man or wife,
to save a pound,
in house or ground,
  ech other weeke?
What more for health,
what more for wealth,
what needeth lesse,
run Jack, helpe Besse,
to staie amis,
not having this,
  far off to seeke?

I do not crave
mo thankes to have,
than given to me
alreadie be,
but this is all
to such as shall
　peruse this booke:
That for my sake,
they gently take,
where ere they finde
against their minde,
when he or she
shall minded be
　therein to looke.

And grant me now,
thou reader thow,
of termes to use,
such choise to chuse,
as may delight
the countrie wight,
　and knowledge bring:
For such doe praise
the countrie phraise,
the countrie acts,
the countrie facts,
the countrie toies,
before the joies
　of anie thing.

Nor looke thou here
that everie shere
of everie verse
I thus reherse
may profit take
or vantage make
　by lessons such:
For here we see
things severall bee,
and there no dike,
but champion like,
and sandie soile,
and claiey toile,
　doe suffer much.

This being waid,
be not afraid
to buie to prove,
to reade with love,
to followe some,
and so to come
　by practise true:
My paine is past,
thou warning hast,
th' experience mine,
the vantage thine,
may give thee choice
to crie or rejoice:
　and thus adue.

Finis T. Tusser.

AN INTRODUCTION
to the Booke of Husbandrie.

Chapter 4

GOOD husbandmen must moile & toile,
   to laie to live by laboured feeld:
Their wives at home must keepe such coile,
   as their like actes may profit yeeld.
      For well they knowe,
      as shaft from bowe,
      or chalke from snowe,
A good round rent their Lords they give,
   and must keepe touch in all their paie:
With credit crackt else for to live,
   or trust to legs and run awaie.

Though fence well kept is one good point,
   and tilth well done, in season due;
Yet needing salve in time to annoint,
   is all in all and needfull true:
      As for the rest,
      thus thinke I best,
      as friend doth gest,
With hand in hand to leade thee foorth   *Ceres,*
   to *Ceres* campe, there to behold   *Goddesse of*
A thousand things as richlie woorth,   *husbandry.*
   as any pearle is woorthie gold.

A PREFACE TO THE BUYER
of this Booke.

Chapter 5

WHAT lookest thou herein to have?
Fine verses thy fansie to please?
Of many my betters that crave,
Looke nothing but rudenes in thease.

What other thing lookest thou then?
Grave sentences many to finde?
Such, Poets have twentie and ten,
Yea thousands contenting the minde.

What looke ye, I praie you shew what?
Termes painted with Rhetorike fine?
Good husbandrie seeketh not that,
Nor ist any meaning of mine.

What lookest thou, speake at the last?
Good lessons for thee and thy wife?
Then keepe them in memorie fast,
To helpe as a comfort to life.

What looke ye for more in my booke?
Points needfull and meete to be knowne?
Then dailie be suer to looke,
To save to be suer thine owne.

THE COMMODITIES OF HUSBANDRIE

Chapter 6

Let house have to fill her,
Let land have to till her.

NO dwellers, what profiteth house for to stand?
What goodnes, unoccupied, bringeth the land?

No labor no bread,
No host we be dead.

No husbandry used, how soone shall we sterve?
House keeping neglected, what comfort to serve?

Ill father no gift,
No knowledge no thrift.

The father an unthrift, what hope to the sonne?
The ruler unskilfull, how quickly undonne?

Chapter 7

As true as thy faith
This riddle thus saith.

I SEEME but a drudge, yet I passe any King   *The praise of*
To such as can use me, great wealth I do bring.   *husbandrie.*
Since Adam first lived, I never did die,
When Noe was shipman, there also was I.
The earth to susteine me, the sea for my fish:
Be readie to pleasure me, as I would wish.
What hath any life, but I helpe to preserve,
What wight without me, but is ready to sterve.

In woodland, in Champion, Citie, or towne
If long I be absent, what falleth not downe?
If long I be present, what goodnes can want?
Though things at my comming were never so scant.
So many as loove me, and use me aright,
With treasure and pleasure, I richly acquite.
Great kings I doe succour, else wrong it would go,
The King of al kings hath appointed it so.

THE DESCRIPTION OF HUSBANDRIE

Chapter 8

OF husband, doth husbandrie challenge that name,
   of husbandrie, husband doth likewise the same:
Where huswife and huswiferie, joineth with thease,
   there wealth in abundance is gotten with ease.

The name of a husband, what is it to saie ?
   of wife and the houshold the band and the staie:
Some husbandlie thriveth that never had wife,
   yet scarce a good husband in goodnes of life.

The husband is he that to labour doth fall,
   the labour of him I doe husbandrie call:
If thrift by that labour be any way caught,
   then is it good husbandrie, else it is naught.

So houshold and housholdrie I doe define,
   for folke and the goodes that in house be of thine:
House keeping to them, as a refuge is set,
   which like as it is, so report it doth get.

Be house or the furniture never so rude,
   of husband and husbandrie, (thus I conclude:)
That huswife and huswiferie, if it be good,
   must pleasure togither as cosins in blood.

THE LADDER TO THRIFT

Chapter 9

TO take thy calling thankfully,
and shun the path to beggery.
To grudge in youth no drudgery,
to come by knowledge perfectly.
To count no travell slaverie,
that brings in penie saverlie.
To folow profit earnestlie
but meddle not with pilferie.
To get by honest practisie,
and keepe thy gettings covertlie.
To lash not out too lashinglie,
for feare of pinching penurie.
To get good plot to occupie,
and store and use it husbandlie.
To shew to landlord curtesie,
and keepe thy covenants orderlie.
To hold that thine is lawfullie,
for stoutnes or for flatterie.
To wed good wife for companie,
and live in wedlock honestlie.
To furnish house with housholdry,
and make provision skilfully.
To joine to wife good familie,
and none to keepe for braverie.
To suffer none live idlelie,
for feare of idle knaverie.
To courage wife in huswiferie,
and use well dooers gentilie.
To keepe no more but needfullie,
and count excesse unsaverie.
To raise betimes the lubberlie,
both snorting Hob and Margerie.
To walke thy pastures usuallie,
to spie ill neighbours subtiltie.
To hate revengement hastilie,
for loosing love and amitie.
To love thy neighbor neighborly,
and shew him no discurtesy.
To answere stranger civilie,
but shew him not thy secresie.
To use no friend deceitfully,
to offer no man villeny.
To learne how foe to pacifie,
but trust him not too trustilie.
To keepe thy touch substanciallie,
and in thy word use constancie.
To make thy bandes advisedly,
& com not bound through suerty.

To meddle not with usurie,
nor lend thy monie foolishlie.
To hate to live in infamie,
through craft, and living shiftingly.
To shun all kinde of treachery,
for treason endeth horribly.
To learne to eschew ill companie,
and such as live dishonestly.
To banish house of blasphemie,
least crosses crosse unluckelie.
To stop mischance, through policy,
for chancing too unhappily.

To beare thy crosses paciently,
for worldly things are slippery.
To laie to keepe from miserie,
age comming on so creepinglie.
To praie to God continuallie,
for aide against thine enimie.
To spend thy Sabboth holilie,
and helpe the needie povertie.
To live in conscience quietly,
and keepe thy selfe from malady.
To ease thy sicknes speedilie,
er helpe be past recoverie.
To seeke to God for remedie,
for witches proove unluckilie.
These be the steps unfainedlie:
to climbe to thrift by husbandrie.

These steps both reach, and teach thee shall:
To come by thrift, to shift withall.

GOOD HUSBANDLIE LESSONS
worthie to be followed of such
as will thrive.

Chapter 10

GOD sendeth and giveth both mouth and the meat,
    and blesseth us al with his benefits great:
Then serve we that God that so richly doth give,
    shew love to our neighbors, and lay for to live.

As bud by appearing betokneth the spring,
    and leafe by her falling the contrarie thing:
So youth bids us labour, to get as we can,
    for age is a burden to laboring man.

A competent living, and honestly had,
    makes such as are godlie both thankfull and glad:
Life never contented, with honest estate,
    lamented is oft, and repented too late.

Count never wel gotten that naughtly is got,
    nor well to account of which honest is not:
Looke long not to prosper, that wayest not this,
    least prospering faileth, and all go amisse.

True wedlock is best, for avoiding of sinne,   *Laie wisely*
    the bed undefiled much honour doth winne: *to marrie.*
Though love be in choosing farre better than gold,
    let love come with somewhat, the better to hold.

Where cooples agree not is ranker and strife,   *Concord*
    where such be together is seldome good life: *bringeth*
Where cooples in wedlcok doe lovelie agree,   *foyson.*
    there foyson remaineth, if wisedome there bee.

Wife and children crave a dwelling.

Who looketh to marrie must laie to keepe house,
 for love may not alway be plaieing with douse:
If children encrease, and no staie of thine owne,
 what afterwards followes is soone to be knowne.

Thee for thrive. Hostisses grudge: nurses crave.

Once charged with children, or likelie to bee,
 give over to sudgerne, that thinkest to thee:
Least grutching of hostis, and craving of nurse,
 be costlie and noisome to thee and thy purse.

Live within thy Tedder.

Good husbands that loveth good houses to keepe
 are oftentimes carefull when other doe sleepe:
To spend as they may, or to stop at the furst,
 for running in danger, or feare of the wurst.

By harvest is ment al thy stock.

Go count with thy cofers, when harvest is in,
 which waie for thy profite, to save or to win:
Of tone of them both, if a saver wee smel,
 house keeping is godlie where ever we dwel.

Be thine own purs bearer.

Sonne, think not thy monie purse bottom to burn,
 but keepe it for profite, to serve thine owne turn:
A foole and his monie be soone at debate,
 which after with sorrow repents him too late.

Good bargaine a dooing, make privie but few,
 in selling, refraine not abrode it to shew:
In making make haste, and awaie to thy pouch,
 in selling no haste, if ye dare it avouch.

Evill landlord.

Good Landlord who findeth, is blessed of God,
 A cumbersome Landlord is husbandmans rod:
He noieth, destroieth, and al to this drift,
 to strip his poore tenant of ferme and of thrift.

Rent corne.

Rent corn who so paieth, (as worldlings wold have,
 so much for an aker) must live as a slave:
Rent corne to be paid, for a reasnable rent,
 at reasnable prises is not to lament.

Once placed for profit, looke never for ease,
  except ye beware of such michers as thease:
Unthiftines, Slouthfulnes, Careles and Rash,
  that trusteth thee headlong to run in the lash.

Foure beggers.

Make monie thy drudge, for to follow thy warke,
  Make wisedome controler, good order thy clarke:
Provision Cater, and skil to be cooke,
  make steward of all, pen, inke, and thy booke.

Thrifts officers.

Make hunger thy sauce, as a medcine for helth,
  make thirst to be butler, as physick for welth:
Make eie to be usher, good usage to have,
  make bolt to be porter, to keepe out a knave.

Thrifts phisicke.

Make husbandrie bailie, abrode to provide,
  make huswiferie dailie at home for to guide:
Make cofer fast locked, thy treasure to keepe,
  make house to be sure, the safer to sleepe.

Thrifts bailie.

Make bandog thy scoutwatch, to barke at a theefe,
  make courage for life to be capitaine cheefe:
Make trapdore thy bulwarke, make bell to begin,
  make gunstone and arrow shew who is within.

Husbandly armors.

The credite of maister, to brothell his man,
  and also of mistresse, to minnekin Nan,
Be causers of opening a number of gaps,
  That letteth in mischiefe and many mishaps.

Theeves to thrift.

Good husband he trudgeth, to bring in the gaines,
  good huswife she drudgeth, refusing no paines:
Though husband at home be to count ye wote what,
  yet huswife within is as needfull as that.

Friends to thrift.

What helpeth in store to have never so much,
  halfe lost by ill usage, ill huswives, and such:
So, twentie lode bushes, cut downe at a clap,
  such heede may be taken, shall stop but a gap.

Enimie to thrift

Sixe noiances to thrift.

A retcheles servant, a mistres that scowles,
   a ravening mastife, and hogs that eate fowles:
A giddie braine maister, and stroyal his knave,
   brings ruling to ruine, and thrift to hir grave.

Inough is a praise.

With some upon Sundaies, their tables doe reeke,
   and halfe the weeke after, their dinners to seeke:
Not often exceeding, but alwaie inough,
   is husbandlie fare, and the guise of the plough.

Ech daie to be feasted, what husbandrie wurse,
   ech daie for to feast, is as ill for the purse:
Yet measurely feasting with neighbors among,
   shal make thee beloved, and live the more long.

Thrifts advises.

Things husbandly handsom let workman contrive,
   but build not for glorie, that thinkest to thrive:
Who fondlie in dooing consumeth his stock,
   in the end for his follie doth get but a mock.

Spoilers to thrift.

Spend none but your owne, howsoever ye spend,
   for bribing and shifting, have seldom good end:
In substance although ye have never so much,
   delight not in parasites, harlots, and such.

Be suretie seldome, (but never for much)
   for feare of purse pennilles hanging by such:
Or Skarborow warning, as ill I beleeve,
   when (sir I arest yee) gets hold of thy sleeve.

Use (*legem pone*) to paie at thy daie,
   but use not (*Oremus*) for often delaie:
Yet (*Præsta quæsumus*) out of a grate,
   Of al other collects, the lender doth hate.

Be pinched by lending, for kiffe nor for kin,
   nor also by spending, by such as come in;
Nor put to thy hand betwixt bark and the tree,
   least through thy owne follie so pinched thou bee.

As lending to neighbour, in time of his neede,
    winnes love of thy neighbour, and credit doth breede,
So never to crave, but to live of thine owne,
    brings comforts a thousand, to many unknowne.

Who living but lends? and be lent to they must;
    else buieng and selling might lie in the dust;
But shameles and craftie, that desperate are,
    make many ful honest the woorser to fare.

At some time to borow, account it no shame,
    if justlie thou keepest thy touch for the same:
Who quick be to borow, and slow be to paie,
    their credit is naught, go they never so gaie.

By shifting and borrowing, who so as lives,
    not well to be thought on, occasion gives:
Then lay to live warily, and wisely to spend,
    for prodigall livers have seldom good end.

Some spareth too late, and a number with him,
    the foole at the bottom, the wise at the brim:
Who careth nor spareth, till spent he hath all,
    Of bobbing, not robbing, be fearefull he shall.

Where welthines floweth, no friendship can lack,
    whom povertie pincheth, hath friendship as slack:
Then happie is he by example that can
    take heede by the fall of a mischieved man.

Who breaketh his credit, or cracketh it twise,
    trust such with a suretie, if ye be wise:
Or if he be angrie, for asking thy due,
    once even, to him afterward, lend not anue.

Account it wel sold that is justlie well paid,
    and count it wel bought that is never denaid:
But yet here is tone, here is tother doth best,
    for buier and seller, for quiet and rest.

Leave Princes affaires undeskanted on,
   and tend to such dooings as stands thee upon:
Feare God, and offend not the Prince nor his lawes,
   and keepe thyselfe out of the Magistrates clawes.

As interest or usurie plaieth the drevil,
   so hilback and filbellie biteth as evil:
Put dicing among them, and docking the dell:
   and by and by after, of beggerie smell.

Thrifts
Auditor.

Once weekelie remember thy charges to cast,
   once monthlie see how thy expences may last:
If quarter declareth too much to be spent,
   for feare of ill yeere take advise of thy rent.

Who orderlie entreth his paiment in booke,
   may orderlie find them againe (if he looke.)
And he that intendeth but once for to paie:
   shall find this in dooing the quietest waie.

In dealing uprightlie this counsel I teach,
   first recken, then write, er to purse yee doe reach,
Then paie and dispatch him, as soone as ye can:
   for lingring is hinderance to many a man.

Have waights, I advise thee, for silver & gold,
   for some be in knaverie now a daies bold:
And for to be sure good monie to pay:
   receive that is currant, as neere as ye may.

Delight not for pleasure two houses to keepe,
   least charge without measure upon thee doe creepe.
And Jankin and Jenikin coosen thee so
   to make thee repent it, er yeere about go.

The stone that is rouling can gather no mosse,
   who often remooveth is sure of losse.
The rich it compelleth to paie for his pride;
   the poore it undooeth on everie side.

The eie of the maister enricheth the hutch,
    the eie of the mistresse availeth as mutch.
Which eie, if it governe, with reason and skil,
    hath servant and service, at pleasure and wil.

Who seeketh revengement of everie wrong,
    in quiet nor safetie continueth long.
So he that of wilfulnes trieth the law,
    shall strive for a coxcome, and thrive as a daw.

To hunters and haukers, take heede what ye saie,
    milde answere with curtesie drives them awaie:
So, where a mans better wil open a gap,
    resist not with rudenes, for feare of mishap.

A man in this world for a churle that is knowne,
    shall hardlie in quiet keepe that is his owne:
Where lowlie and such as of curtesie smels,
    finds favor and friendship where ever he dwels.

Keepe truelie thy Saboth, the better to speed,
    Keepe servant from gadding, but when it is need.
Keepe fishdaie and fasting daie, as they doe fal:
    what custome thou keepest, let others keepe al.

Though some in their tithing be slack or too bold,
    be thou unto Godward not that waie too cold:
Evill conscience grudgeth, and yet we doe see
    ill tithers ill thrivers most commonlie bee.

Paie weekelie thy workman, his houshold to feed,
    paie quarterlie servants, to buie as they need:
Give garment to such as deserve and no mo,
    least thou and thy wife without garment doe go.

Beware raskabilia, slothfull to wurke,
    purloiners and filchers, that loveth to lurke.
Away with such lubbers, so loth to take paine,
    that roules in expences, but never no gaine.

Good wife, and good children, are worthie to eate,
  good servant, good laborer, earneth their meate:
Good friend, and good neighbor, that fellowlie gest,
  with hartilie welcome, should have of the best.

Depart not with al that thou hast to thy childe,
  much lesse unto other, for being beguilde:
Least, if thou wouldst gladlie possesse it agen,
  looke for to come by it thou wottest not when.

The greatest preferment that childe we can give,
  is learning and nurture, to traine him to live:
Which who so it wanteth, though left as a squier,
  consumeth to nothing, as block in the fier.

When God hath so blest thee, as able to live,
  and thou hast to rest thee, and able to give,
Lament thy offences, serve God for amends,
  make soule to be readie when God for it sends.

Send fruites of thy faith to heaven aforehand,
  for mercie here dooing, God blesseth thy land:
He maketh thy store with his blessing to swim,
  and after, thy soule to be blessed with him.

Some lay to get riches by sea and by land,
  and ventreth his life in his enimies hand:
And setteth his soule upon sixe or on seaven,
  not fearing nor caring for hell nor for heaven.

Some pincheth, and spareth, and pineth his life,
  to cofer up bags for to leave to his wife:
And she (when he dieth) sets open the chest,
  for such as can sooth hir and all away wrest.

Good husband, preventing the frailnes of some,
  takes part of Gods benefits, as they doo come,
And leaveth to wife and his children the rest,
  each one his owne part, as he thinketh it best.

These lessons approoved, if wiselie ye note,
may save and avantage ye many a grote.
Which if ye can follow, occasion found,
then everie lesson may save ye a pound.

AN Habitation inforced better late than never,
upon these words Sit downe Robin and rest thee.

Chapter 11

MY friend, if cause doth wrest thee,
Ere follie hath much opprest thee:
Farre from acquaintance kest thee,
Where countrie may digest thee,
Let wood and water request thee,
In good corne soile to nest thee,
Where pasture and meade may brest thee,
And healthsom aire invest thee.
Though envie shall detest thee,
Let that no whit molest thee,
Thanke God, that so hath blest thee,
And sit downe Robin & rest thee.

THE FERMERS DAILIE DIET

Chapter 12

A PLOT set downe, for fermers quiet,
as time requires, to frame his diet:
With sometime fish, and sometime fast,
that houshold store may longer last.

Lent.	Let Lent well kept offend not thee,
	   for March and Aprill breeders bee:
	Spend herring first, save saltfish last,
	   for saltfish is good, when Lent is past.

Easter.	When Easter comes, who knowes not than,
	   that Veale and Bakon is the man:
	And Martilmas beefe doth beare good tack,
	   when countrie folke doe dainties lack.

Midsommer.	When Mackrell ceaseth from the seas, John Baptist brings grassebeefe and pease.
Mihelmas.	Fresh herring plentie, Mihell brings, with fatted Crones, and such old things.
Hallomas.	All Saints doe laie for porke and souse, for sprats and spurlings for their house.
Christmas.	At Christmas play and make good cheere, for Christmas comes but once a yeere.
A caveat.	Though some then doe, as doe they would, let thriftie doe, as doe they should.
Fasting.	For causes good, so many waies, keepe Embrings wel, and fasting daies:
Fish daies.	What lawe commands, we ought to obay, for Friday, Saturne, and Wednesday.
A thing needful.	The land doth will, the sea doth wish, spare sometime flesh, and feede of fish.
The last remedie.	Where fish is scant, and fruit of trees, Supplie that want with butter and cheese.

                                                q. Tusser.

A description of the properties of windes
all the times of the yeere.

Chapter 13

North winds send haile, South winds bring raine,   *In winter.*
East winds we bewail, West winds blow amaine:
North east is too cold, South east not too warme,
North west is too bold, South west doth no harme.

The north is a noyer to grasse of all suites,   *At the spring.*
The east a destroyer to herbe and all fruites:
The south with his showers refresheth the corne,   *Sommer.*
The west to all flowers may not be forborne.

The West, as a father, all goodnes doth bring,   *Autumne.*
The East, a forbearer, no manner of thing:
The South, as unkind, draweth sicknesse too neere,
The North, as a friend, maketh all againe cleere.

With temperate winde we be blessed of God,   *God is the*
With tempest we finde we are beat with his rod: *governer of*
All power we knowe to remaine in his hand,   *winde and*
How ever winde blowe, by sea or by land.   *weather.*

Though windes doe rage, as windes were wood,
And cause spring tydes to raise great flood,
And loftie ships leave anker in mud,
Bereafing many of life and of blud;

Yet true it is, as cow chawes cud,
And trees at spring doe yeeld forth bud,
Except winde stands as never it stood,
It is an ill winde turnes none to good.

OF THE PLANETS

Chapter 14

AS huswives are teached, in stead of a clock,
   how winter nights passeth, by crowing of cock;
So here by the Planets, as far as I dare,
   some lessons I leave for the husbandmans share.

Of the rising and going down of the sun.

If day star appeareth, day comfort is ny,
   If sunne be at south, it is noone by and by:
If sunne be at westward, it setteth anon,
   If sunne be at setting, the day is soone gon.

Of the Moone changing.

Moone changed, keepes closet three daies as a Queene,
   er she in hir prime will of any be seene:
If great she appereth, it showreth out,
   If small she appereth, it signifieth drout.
At change or at full, come it late or else soone,
   maine sea is at highest, at midnight and noone:
But yet in the creekes it is later high flood,
   through farnesse of running, by reason as good.

Of flowing and ebbing to such as be verie sick.

Tyde flowing is feared, for many a thing,
   great danger to such as be sick it doth bring:
Sea eb by long ebbing some respit doth give,
   and sendeth good comfort to such as shal live.

SEPTEMBERS ABSTRACT

Chapter 15

Now enter John,
old fermer is gon.

What champion useth,
that woodland refuseth.

Good ferme now take,
keepe still, or forsake.

What helpes to revive
the thriving to thrive.

Plough, fence, and store
aught else before.

By tits and such
few gaineth much.

Horse strong and light
soone charges quite.
Light head and purse,
what lightnes wurse.

Who goeth a borrowing,
goeth a sorrowing.
Few lends (but fooles)
their working tooles.

Greene rie have some,
er Mihelmas come.

Grant soile hir lust,
sowe rie in the dust.

Cleane rie that sowes,
the better crop mowes.

Mix rie aright,
with wheat that is whight.

See corne sowen in,
too thick nor too thin.
For want of seede,
land yeeldeth weede.

With sling or bowe,
keepe corne from Crowe.

Trench hedge and forrow,
that water may thorow.
Deepe dike saves much,
from drovers and such.

Amend marsh wall,
Crab holes and all.

Geld bulles and rams,
sewe ponds, amend dams.
Sell webster thy wull,
fruite gather, grapes pull.
For fear of drabs,
go gather thy crabs.

Plucke fruite to last,
when Mihell is past.

Forget it not,
fruit brused will rot.
Light ladder and long
doth tree least wrong.
Go gather with skill,
and gather that will.

Drive hive, good conie,
for waxe and for honie.
No driving of hive,
till yeeres past five.

Good dwelling give bee,
or hence goes shee.

Put bore in stie,
for Hallontide nie.

With bore (good Cisse)
let naught be amisse.

Karle hempe, left greene,
now pluck up cleene.
Drowne hemp as ye need,
once had out his seed.
I pray thee (good Kit)
drowne hempe in pit.

Of al the rest,
white hempe is best.
Let skilfull be gotten
least hempe proove rotten.

Set strawberies, wife,
I love them for life.

Plant Respe and rose,
and such as those.

Goe gather up mast,
er time be past.
Mast fats up swine,
Mast kils up kine.

Let hogs be roong,
both old and yoong.

No mast upon oke,
no longer unyoke.
If hog doe crie,
give eare and eie.

Hogs haunting corne
may not be borne.

Good neighbour thow
good custome alow,
No scaring with dog,
whilst mast is for hog.

Get home with the brake,
to brue with and bake,
To cover the shed
drie over the hed,
To lie under cow,
to rot under mow,
To serve to burne,
for many a turne.

To sawpit drawe
boord log, to sawe.
Let timber be haile,
least profit doe quaile.
Such boord and pale
is readie sale.

Sawne slab let lie,
for stable and stie,
sawe dust spred thick,
makes alley trick.

Keepe safe thy fence,
scare breakhedge thence.
A drab and a knave
will prowle to have.

Marke winde and moone,
at midnight and noone.
Some rigs thy plow,
some milks thy cow.

Red cur or black,
few prowlers lack.
Some steale, some pilch,
some all away filch,
Mark losses with greefe,
through prowling theefe.

*Thus endeth Septembers
abstract, agreeing with
Septembers husbandrie.*

OTHER SHORT REMEMBRANCES

Now friend, as ye wish,
goe sever thy fish:
When friend shall come,
to be sure of some.

Thy ponds renew,
put eeles in stew,
To leeve till Lent,
and then to be spent.

Set privie or prim,
set boxe like him.
Set Giloflowers all,
that growes on the wall.

Set herbes some more,
for winter store.
Sowe seedes for pot,
for flowers sowe not.

*Here ends Septembers
short remembrances*

SEPTEMBERS HUSBANDRIE

Chapter 15

September blowe soft,   *Forgotten, month past,*
Till fruite be in loft.   *Doe now at the last.*

AT Mihelmas lightly new fermer comes in,
   new husbandrie forceth him new to begin:
Old fermer, still taking the time to him given,
   makes August to last untill Mihelmas even.

New fermer may enter (as champions say)
   on all that is fallow, at Lent ladie day:
In woodland, old fermer to that will not yeeld,
   for loosing of pasture, and feede of his feeld.

Ferme take or give over.
Provide against Mihelmas, bargaine to make,
   for ferme to give over, to keepe or to take:
In dooing of either, let wit beare a stroke,
   for bueing or selling of a pig in a poke.

Twelve good properties.
Good ferme and well stored, good housing and drie,
   good corne and good dairie, good market and nie:
Good shepheard, good tilman, good Jack and good Gil,
   makes husband and huswife their cofers to fil.

Have ever a good fence.
Let pasture be stored, and fenced about,
   and tillage set forward, as needeth without:
Before ye doe open your purse to begin,
   with anything dooing for fancie within.

Best cattle most profit.
No storing of pasture with baggedglie tit,
   with ragged, with aged, and evil athit:
Let carren and barren be shifted awaie,
   for best is the best, whatsoever ye paie.

Horse, Oxen, plough, tumbrel, cart, waggon, & waine,   *Strong and*
   the lighter and stronger, the greater thy gaine.   *light.*
The soile and the seede, with the sheafe and the purse,
   the lighter in substance, for profite the wurse.

To borow to daie and to-morrow to mis,   *Hate borow-*
   for lender and borower, noiance it is: *ing.*
Then have of thine owne, without lending unspilt,
   what followeth needfull, here learne if thou wilt.

A DIGRESSION TO HUSBANDLIE FURNITURE

Barne locked, gofe ladder, short pitchforke and long,   *Barne furni-*
   flaile, strawforke and rake, with a fan that is strong: *ture.*
Wing, cartnave and bushel, peck, strike readie hand,
   get casting sholve, broome, and a sack with a band.

A stable wel planked, with key and a lock,   *Stable furni-*
   walles stronglie wel lyned, to beare off a knock: *ture.*
A rack and a manger, good litter and haie,
   sweete chaffe and some provender everie daie.

A pitchfork, a doongfork, seeve, skep and a bin,
   a broome and a paile to put water therein:
A handbarow, wheelebarow, sholve and a spade,
   a currie combe, mainecombe, and whip for a Jade.

A buttrice and pincers, a hammer and naile,
   an aperne and siszers for head and for taile:
Hole bridle and saddle, whit lether and nall,
   with collers and harneis, for thiller and all.

A panel and wantey, packsaddle and ped,
   A line to fetch litter, and halters for hed.
With crotchis and pinnes, to hang trinkets theron,
   and stable fast chained, that nothing be gon.

Cart furniture.	Strong exeltred cart, that is clouted and shod,
	   cart ladder and wimble, with percer and pod:
	Wheele ladder for harvest, light pitchfork and tough,
	   shave, whiplash wel knotted, and cartrope ynough.

A Coeme is halfe a quarter.	Ten sacks, whereof everie one holdeth a coome,
	   a pulling hooke handsome, for bushes and broome:
	Light tumbrel and doong crone, for easing sir wag,
	   sholve, pickax, and mattock, with bottle and bag.

Husbandry tooles.	A grinstone, a whetstone, a hatchet and bil,
	   with hamer and english naile, sorted with skil:
	A frower of iron, for cleaving of lath,
	   with roule for a sawpit, good husbandrie hath.

	A short saw and long saw, to cut a too logs,
	   an ax and a nads, to make troffe for thy hogs:
	A Dovercourt beetle, and wedges with steele,
	   strong lever to raise up the block fro the wheele.

Plough furniture.	Two ploughs and a plough chein, ij culters, iij shares,
	   with ground cloutes & side clouts for soile that so tares:
	With ox bowes and oxyokes, and other things mo,
	   for oxteeme and horseteeme, in plough for to go.

	A plough beetle, ploughstaff, to further the plough,
	   great clod to a sunder that breaketh so rough;
	A sled for a plough, and another for blocks,
	   for chimney in winter, to burne up their docks.

	Sedge collers for ploughhorse, for lightnes of neck,
	   good seede and good sower, and also seede peck:
	Strong oxen and horses, wel shod and wel clad,
	   wel meated and used, for making thee sad.

	A barlie rake toothed, with yron and steele,
	   like paier of harrowes, and roler doth weele:
	A sling for a moether, a bowe for a boy.
	   a whip for a carter, is hoigh de la roy.

A brush sithe and grasse sithe, with rifle to stand,   *Harvest*
   a cradle for barlie, with rubstone and sand: *tooles.*
Sharpe sikle and weeding hooke, haie fork and rake,
   a meake for the pease, and to swinge up the brake.

Short rakes for to gather up barlie to binde,
   and greater to rake up such leavings behinde:
A rake for to hale up the fitchis that lie,
   a pike for to pike them up handsom to drie.

A skuttle or skreine, to rid soile fro the corne,
   and sharing sheares readie for sheepe to be shorne:
A fork and a hooke, to be tampring in claie,
   a lath hammer, trowel, a hod, or a traie.

Strong yoke for a hog, with a twicher and rings,
   with tar in a tarpot, for dangerous things:
A sheepe marke, a tar kettle, little or mitch,
   two pottles of tar to a pottle of pitch.

Long ladder to hang al along by the wal,
   to reach for a neede to the top of thy hal:
Beame, scales, with the weights, that be sealed and true,
   sharp moulspare with barbs, that the mowles do so rue.

Sharpe cutting spade, for the deviding of mow,
   with skuppat and skavel, that marsh men alow:
A sickle to cut with, a didall and crome
   for draining of ditches, that noies thee at home.

A clavestock and rabetstock, carpenters crave,
   and seasoned timber, for pinwood to have:
A Jack for to saw upon fewell for fier,
   for sparing of firewood, and sticks fro the mier.

Soles, fetters, and shackles, with horselock and pad,
   a cow house for winter, so meete to be had:
A stie for a bore, and a hogscote for hog,
   a roost for thy hennes, and a couch for thy dog.

      Here endeth husbandlie furniture.

Sowing Thresh seed and to fanning, September doth crie,
of rie.   get plough to the field, and be sowing of rie:
To harrow the rydgis, er ever ye strike,
   is one peece of husbandrie Suffolk doth like.

Sowe timely thy whitewheat, sowe rie in the dust,
   let seede have his longing, let soile have hir lust:
Let rie be partaker of Mihelmas spring,
   to beare out the hardnes that winter doth bring.

Myslen. Some mixeth to miller the rie with the wheat,
   Temmes lofe on his table to have for to eate:
But sowe it not mixed, to growe so on land,
   least rie tarie wheat, till it shed as it stand.

If soile doe desire to have rie with the wheat,
   by growing togither, for safetie more great,
Let white wheat be ton, be it deere, be it cheape,
   the sooner to ripe, for the sickle to reape.

Sowing. Though beanes be in sowing but scattered in,
   yet wheat, rie, and peason, I love not too thin:
Sowe barlie and dredge, with a plentifull hand,
   least weede, steed of seede, over groweth thy land.

No sooner a sowing, but out by and by,
   with mother or boy that Alarum can cry:
Keeping of And let them be armed with sling or with bowe,
crowes.   to skare away piggen, the rooke and the crowe.

Water fur- Seed sowen, draw a forrough, the water to draine,
rough.   and dike up such ends as in harmes doe remaine:
For driving of cattell or roving that waie,
   which being prevented, ye hinder their praie.

Amend Saint Mihel doth bid thee amend the marsh wal,
marsh   the brecke and the crab hole, the foreland and al:
walles. One noble in season bestowed theron,
   may save thee a hundred er winter be gon.

Now geld with the gelder the ram and the bul,
    sew ponds, amend dammes, and sel webster thy wul:
 Out fruit go and gather, but not in the deaw,
    with crab and the wal nut, for feare of a shreaw
	Gelding of rams.

The Moone in the wane, gather fruit for to last,
    but winter fruit gather when Mihel is past:
 Though michers that love not to buy nor to crave,
    makes some gather sooner, else few for to have.
	Gathering of fruit.

Fruit gathred too timely wil taste of the wood,
    wil shrink and be bitter, and seldome proove good:
 So fruit that is shaken, or beat off a tree,
    with brusing in falling, soone faultie wil bee.
	Too early gathering is not best.

Now burne up the bees that ye mind for to drive,
    at Midsomer drive them and save them alive:
 Place hive in good ayer, set southly and warme,
    and take in due season wax, honie, and swarme.
	Driving of hives.

Set hive on a plank, (not too low by the ground)
    where herbe with the flowers may compas it round:
 And boordes to defend it from north and north east,
    from showers and rubbish, from vermin and beast.
	Preserving of bees.

At Mihelmas safely go stie up thy Bore,
    least straying abrode, ye doo see him no more:
 The sooner the better for Halontide nie,
    and better he brawneth if hard he doo lie.
	Stie up the bore.

Shift bore (for il aire) as best ye do thinke,
    and twise a day give him fresh vittle and drinke:
 And diligent Cislye, my dayrie good wench,
    make cleanly his cabben, for measling and stench.	

Now pluck up thy hempe, and go beat out the seed,
    and afterward water it as ye see need:
 But not in the river where cattle should drinke,
    for poisoning them and the people with stinke.
	Gathering of winter hempe.

Whitest hempe best sold.

Hempe huswifely used lookes cleerely and bright,
   and selleth it selfe by the colour so whight:
Some useth to water it, some do it not,
   be skilful in dooing, for feare it do rot.

Setting of strawberries & roses, &c.

Wife, into thy garden, and set me a plot,
   with strawbery rootes, of the best to be got:
Such growing abroade, among thornes in the wood,
   wel chosen and picked proove excellent good.

Gooseberies & Respis.

The Barbery, Respis, and Goosebery too,
   looke now to be planted as other things doo:
The Goosebery, Respis, and Roses, al three,
   with Strawberies under them trimly agree.

Gathering of mast.

To gather some mast, it shal stand thee upon,
   with servant and children, er mast be al gon:
Some left among bushes shal pleasure thy swine,
   for feare of a mischiefe keepe acrons fro kine.

Rooting of hogs.

For rooting of pasture ring hog ye had neede,
   which being wel ringled the better do feede:
Though yong with their elders wil lightly keepe best,
   yet spare not to ringle both great and the rest.

Yoking of swine.

Yoke seldom thy swine while the shacktime doth last,
   for divers misfortunes that happen too fast:
Or if ye do fancie whole eare of the hog,
   give eie to il neighbour and eare to his dog.

Hunting of hogs.

Keepe hog I advise thee from medow and corne,
   for out aloude crying that ere he was borne:
Such lawles, so haunting, both often and long,
   if dog set him chaunting he doth thee no wrong.

Where love among neighbors do beare any stroke,
   whiles shacktime indureth men use not to yoke:

Ringling of hogs.

Yet surely ringling is needeful and good,
   til frost do enuite them to brakes in the wood.

Get home with thy brakes, er an sommer be gon,   *Carriage of*
   for teddered cattle to sit there upon: *brakes.*
To cover thy hovel, to brewe and to bake,
   to lie in the bottome, where hovel ye make.

Now sawe out thy timber, for boord and for pale,   *Sawe out*
   to have it unshaken, and ready to sale: *thy timber.*
Bestowe it and stick it, and lay it aright,
   to find it in March, to be ready in plight.

Save slab of thy timber for stable and stie,   *Slabs of*
   for horse and for hog the more clenly to lie: *timber.*
Save sawe dust, and brick dust, and ashes so fine,
   for alley to walke in, with neighbour of thine.

Keepe safely and warely thine uttermost fence,   *Hedge*
   with ope gap and brake hedge do seldome dispence: *breakers.*
Such runabout prowlers, by night and by day,
   see punished justly for prowling away.

At noone if it bloweth, at night if it shine,   *Learne to know:*
   out trudgeth Hew make shift, with hooke & with line: *Hewe*
Whiles Gillet, his blouse, is a milking thy cow,   *prowler.*
   Sir Hew is a rigging thy gate or the plow.

Such walke with a black or a red little cur,   *Black or*
   that open wil quickly, if anything stur;   *red dogs.*
Then squatteth the master, or trudgeth away,
   and after dog runneth as fast as he may.

Some prowleth for fewel, and some away rig
   fat goose, and the capon, duck, hen, and the pig:
Some prowleth for acornes, to fat up their swine,
   for corne and for apples, and al that is thine.

Thus endeth Septembers husbandrie.

OCTOBERS ABSTRACT

Chapter 16

LAY drie up and round,
for barlie thy ground.

Too late doth kill,
too soone is as ill.

Maides little and great,
pick cleane seede wheat.
Good ground doth crave
choice seede to have.
Flaies lustily thwack,
least plough seede lack.

Seede first go fetch,
for edish or etch,
Soile perfectly knowe,
er edish ye sowe.

White wheat, if ye please,
sowe now upon pease.
Sowe first the best,
and then the rest.

Who soweth in raine,
hath weed to his paine.
But worse shall he speed,
that soweth ill seed.

Now, better than later,
draw furrow for water.

Keepe crowes, good sonne,
see fencing be donne.

Each soile no vaine
for everie graine.
Though soile be but bad,
some corne may be had.

Naught prove, naught crave,
naught venter, naught have.

One crop and away,
some countrie may say.

All gravell and sand,
is not the best land.
A rottenly mould
is land woorth gould.

Why wheat is smitten
good lesson is written.

The judgement of some
how thistles doe come.

A judgment right,
of land in plight.
Land, all forlorne,
not good for corne.

Land barren doth beare
small strawe, short eare.

Here maist thou reede
for soile what seede.

Tis tride ery hower,
best graine most flower.

Grosse corne much bran
the baker doth ban.

What croppers bee
here learne to see.

Few after crop much,
but noddies and such.

Som woodland may crake,
three crops he may take.

First barlie, then pease,
then wheat, if ye please.

Two crops and away,
must champion say.

Where barlie did growe,
laie wheat to sowe.
Yet better I thinke,
sowe pease after drinke.
And then, if ye please,
sowe wheat after pease.

What champion knowes
that custome showes.

First barlie er rie,
then pease by and by.
Then fallow for wheat,
is husbandrie great.

A remedie sent,
where pease lack vent.
Fat peasefed swine
for drover is fine.

Each divers soile
hath divers toile.

Some countries use
that some refuse.

For wheat ill land,
where water doth stand.
Sowe pease or dredge
belowe in that redge.

Sowe acornes to proove
that timber doe loove.

Sowe hastings now,
if land it alow.

Learne soone to get
a good quickset.

For feare of the wurst
make fat away furst.

Fat that no more
ye keepe for store.

Hide carren in grave,
lesse noiance to have.

Hog measeled kill,
for flemming that will.

With peasebolt and brake
some brew and bake.

Old corne worth gold,
so kept as it shold.

Much profit is rept,
by sloes well kept.

Keepe sloes upon bow,
for flixe of thy cow.

Of vergis be sure,
poore cattel to cure.

Thus endeth Octobers abstract, agreeing with Octobers husbandrie.

OTHER SHORT
REMEMBRANCES

Cisse, have an eie
to bore in the stie.
By malt ill kept,
small profit is rept.

Friend, ringle thy hog,
for feare of a dog.
Rie straw up stack,
least Thacker doe lack.

Wheat straw drie save,
for cattell to have.
Wheat chaffe lay up drie,
in safetie to lie.

Make handsome a bin,
for chaffe to lie in.

(Seede thresht) thou shalt
thresh barlie to malt.
Cut bushes to hedge,
fence medow and redge.

Stamp crabs that may,
for rotting away.
Make vergis and perie,
sowe kirnell and berie.

Now gather up fruite,
of everie suite.
Marsh wall too slight,
strength now, or god night.

Mend wals of mud,
for now it is good.
Where soile is of sand,
quick set out of hand.

To plots not full
ad bremble and hull.
For set no bar
whilst month hath an R.
Like note thou shalt
for making of malt.
Brew now to last
till winter be past.

Here ends Octobers short remembrances.

OCTOBERS HUSBANDRIE

Chapter 17

October good blast, Forgotten month past,
To blowe the hog mast. Doe now at the last.

NOW lay up thy barley land, drie as ye can,
   when ever ye sowe it so looke for it than:
Get daily aforehand, be never behinde;
   least winter preventing do alter thy minde.

Laie up barlie land.

Who laieth up fallow too soone or too wet,
   with noiances many doth barley beset.
For weede and the water so soketh and sucks,
   that goodnes from either it utterly plucks.

Greene rie in September when timely thou hast,
   October for wheat sowing calleth as fast.
If weather will suffer, this counsell I give,
   Leave sowing of wheat before Hallomas eve.

Wheat sowing.

Where wheat upon edish ye mind to bestowe,
   let that be the first of the wheat ye do sowe:
He seemeth to hart it and comfort to bring,
   that giveth it comfort of Mihelmas spring.

Sowe edish betimes.

White wheat upon peaseetch doth grow as he wold,
   but fallow is best, if we did as we shold:
Yet where, how, and when, ye entend to begin,
   let ever the finest be first sowen in.

Best wheat first sowen.

Who soweth in raine, he shall reape it with teares,
   who soweth in harmes, he is ever in feares,
Who soweth ill seede or defraudeth his land,
   hath eie sore abroode, with a coresie at hand.

Seede husbandly sowen, water furrow thy ground,
   that raine when it commeth may run away round,
Then stir about Nicoll, with arrow and bowe,
   take penie for killing of everie crowe.

A digression to the usage of divers countries,
   CONCERNING TILLAGE

Each soile hath no liking of everie graine,
   nor barlie and wheat is for everie vaine:
Yet knowe I no countrie so barren of soile
   but some kind of corne may be gotten with toile.

In Brantham, where rie but no barlie did growe,
   good barlie I had, as a meany did knowe:
Five seame of an aker I truely was paid,
   for thirtie lode muck of each aker so laid.

In Suffolke againe, where as wheat never grew,
   good husbandrie used good wheat land I knew:
This Proverbe experience long ago gave,
   that nothing who practiseth nothing shall have.

As gravell and sand is for rie and not wheat,
   (or yeeldeth hir burden to tone the more great,)
So peason and barlie delight not in sand,
   but rather in claie or in rottener land.

Wheat somtime is steelie or burnt as it growes,
   for pride or for povertie practise so knowes.
Too lustie of courage for wheat doth not well,
   nor after sir peeler he looveth to dwell.

Much wetnes, hog rooting, and land out of hart,
   makes thistles a number foorthwith to upstart.
If thistles so growing proove lustie and long,
   it signifieth land to be hartie and strong.

As land full of tilth and in hartie good plight,
   yeelds blade to a length and encreaseth in might,
So crop upon crop, upon whose courage we doubt,
   yeelds blade for a brag, but it holdeth not out.

The straw and the eare to have bignes and length,
   betokeneth land to be good and in strength.
If eare be but short, and the strawe be but small,
   it signifieth barenes and barren withall.

White wheat or else red, red rivet or whight,
   far passeth all other, for land that is light.
White pollard or red, that so richly is set,
   for land that is heavie is best ye can get.

Maine wheat that is mixed with white and with red
   is next to the best in the market mans hed:
So Turkey or Purkey wheat many doe love,
   because it is flourie, as others above.

Graie wheat is the grosest, yet good for the clay,
   though woorst for the market, as fermer may say.
Much like unto rie be his properties found,
   coorse flower, much bran, and a peeler of ground.

Otes, rie, or else barlie, and wheat that is gray,
   brings land out of comfort, and soone to decay:
One after another, no comfort betweene,
   is crop upon crop, as will quickly be seene.

Still crop upon crop many fermers do take,   *Crop upon*
   and reape little profit for greedines sake.   *crop.*
Though breadcorne & drinkcorn such croppers do stand:
   count peason or brank, as a comfort to land.

Good land that is severall, crops may have three,
   in champion countrie it may not so bee:
Ton taketh his season, as commoners may,
   the tother with reason may otherwise say.

Some useth at first a good fallow to make,
   to sowe thereon barlie, the better to take.
Next that to sowe pease, and of that to sowe wheat,
   then fallow againe, or lie lay for thy neat.

First rie, and then barlie, the champion saies,
   or wheat before barlie be champion waies:
But drinke before bread corne with Middlesex men,
   then lay on more compas, and fallow agen.

Where barlie ye sowe, after rie or else wheat,
   if land be unlustie, the crop is not great,
So lose ye your cost, to your coresie and smart,
   and land (overburdened) is cleane out of hart.

Exceptions take of the champion land,
   from lieng alonge from that at thy hand.
(Just by) ye may comfort with compas at will,
   far off ye must comfort with favor and skill.

Where rie or else wheat either barlie ye sowe,
   let codware be next, thereupon for to growe:
Thus having two crops, whereof codware is ton,
   thou hast the lesse neede, to lay cost thereupon.

Some far fro the market delight not in pease,
   for that ery chapman they seeme not to please.
If vent of the market place serve thee not well,
   set hogs up a fatting, to drover to sell.

Two crops of a fallow enricheth the plough,
   though tone be of pease, it is land good ynough:
One crop and a fallow some soile will abide,
   where if ye go furder lay profit aside.

Where peason ye had and a fallow thereon,
   sowe wheat ye may well without doong thereupon:
New broken upland, or with water opprest,
   or over much doonged, for wheat is not best.

Where water all winter annoieth too much,
   bestowe not thy wheat upon land that is such:
But rather sowe otes, or else bullimong there,
   gray peason, or runcivals, fitches, or tere.

Sowe acornes ye owners, that timber doe loove,   *Sowing of*
   sowe hawe and rie with them the better to proove;   *acorns.*
If cattel or cunnie may enter to crop,
   yong oke is in daunger of loosing his top.

Who pescods delighteth to have with the furst,   *Sowing of*
   if now he do sowe them, I thinke it not wurst.   *Hastings or*
The greener thy peason and warmer the roome,   *fullams.*
   more lusty the layer, more plenty they come.

Go plow up or delve up, advised with skill,
   the bredth of a ridge, and in length as you will.
Where speedy quickset for a fence ye wil drawe,
   to sowe in the seede of the bremble and hawe.

Through plenty of acornes, the porkling to fat,   *A disease in*
   not taken in season, may perish by that,   *fat hogs.*
If ratling or swelling get once to the throte,
   thou loosest thy porkling, a crowne to a grote.

What ever thing fat is, againe if it fall,   *Not to fat*
   thou ventrest the thing and the fatnes withall,   *for rearing.*
The fatter the better, to sell or to kil,
   but not to continue, make proofe if ye wil.

What ever thing dieth, go burie or burne,   *Burieng of*
   for tainting of ground, or a woorser il turne.   *dead cattell.*
Such pestilent smell of a carrenly thing,
   to cattle and people great peril may bring.

Thy measeled bacon, hog, sow, or thy bore,   *Measeled*
   shut up for to heale, for infecting thy store: *hogs.*
Or kill it for bacon, or sowce it to sell,
   for Flemming, that loves it so deintily well.

Strawwisps and peas-bolts.

With strawisp and peasebolt, with ferne & the brake,
   for sparing of fewel, some brewe and do bake,
And heateth their copper, for seething of graines:
   good servant rewarded, refuseth no paines.

Olde wheat better than new.

Good breadcorne and drinkcorne, full xx weekes kept,
   is better then new, that at harvest is rept:
But foisty the breadcorne and bowd eaten malt,
   for health or for profit, find noysome thou shalt.

By thend of October, go gather up sloes,
   have thou in a readines plentie of thoes,
And keepe them in bedstraw, or still on the bow,
   to staie both the flixe of thyselfe and thy cow.

A medicen for the cow flixe.

Seeith water and plump therein plenty of sloes,
   mix chalke that is dried in powder with thoes.
Which so, if ye give, with the water and chalke,
   thou makest the laxe fro thy cow away walke.

Be sure of vergis (a gallond at least)
   so good for the kitchen, so needfull for beast,
It helpeth thy cattel, so feeble and faint,
   if timely such cattle with it thou acquaint.

      Thus endeth Octobers husbandrie.

NOVEMBERS ABSTRACT

Chapter 18

LET hog once fat,
loose nothing of that.
When mast is gon,
hog falleth anon,
Still fat up some,
till Shroftide come.
Now porke and souse,
beares tack in house.

Pur barlie to malting,
lay flitches a salting.
Through follie too beastlie
much bacon is reastie.

Some winnow, some fan,
some cast that can.
In casting provide,
for seede lay aside.

Thresh barlie thou shalt,
for chapman to malt.
Else thresh no more
but for thy store.

Till March thresh wheat,
but as ye doo eat,
Least baker forsake it
if foystines take it.

No chaffe in bin,
makes horse looke thin,

Sowe hastings now,
that hastings alow.

They buie it full deere,
in winter that reere.

Few fowles, lesse swine,
rere now, friend mine.

What losse, what sturs,
through ravening curs.

Make Martilmas beefe,
deere meate is a theefe.

Set garlike and pease,
saint Edmond to please.

When raine takes place,
to threshing apace.

Mad braine, too rough,
marres all at plough.
With flaile and whips,
fat hen short skips.

Some threshing by taske,
will steale and not aske:
Such thresher at night
walkes seldom home light.

Some corne away lag
in bottle and bag.
Some steales, for a jest,
egges out of the nest.

Lay stover up drie
in order to lie.
Poore bullock doth crave
fresh straw to have.

Make weekly up flower,
though threshers do lower:
Lay graine in loft
and turne it oft.

For muck, regard,
make cleane foule yard.
Lay straw to rot,
in watrie plot.

Hedlond up plow,
for compas ynow.

For herbes good store,
trench garden more.

At midnight trie
foule privies to fie.

Rid chimney of soot,
from top to the foot.

In stable, put now
thy horses for plow.

Good horsekeeper will
laie muck upon hill.

Cut molehils that stand
so thick upon land.

*Thus endeth Novembers
abstract, agreeing with
Novembers husbandrie.*

OTHER SHORT
REMEMBRANCES

Get pole, boy mine,
beate hawes to swine.
Drive hog to the wood,
brake rootes be good.

For mischiefe that falles,
looke well to marsh walles.
Drie laier get neate,
and plentie of meate.

Curst cattel that nurteth,
poore wennel soon hurteth.
Good neighbour mine,
ring well thy swine.

Such winter may serve,
hog ringled will sterve.
In frost keepe dog
from hunting of hog.

Here ends Novembers short remembrances.

NOVEMBERS HUSBANDRIE

Chapter 19

November take flaile, Forgotten month past,
Let ship no more saile. Doe now at the last.

AT Hallontide, slaughter time entereth in, — *Slaughter time.*
   and then doth the husbandmans feasting begin:
From thence unto shroftide kill now and then some,
   their offal for houshold the better wil come.

Thy dredge and thy barley go thresh out to malt, — *Dredge is otes and barlie.*
   let malster be cunning, else lose it thou shalt:
Thencrease of a seame is a bushel for store,
   bad else is the barley, or huswife much more.

Some useth to winnow, some useth to fan, — *Winnowing, fanning, and casting.*
   some useth to cast it as cleane as they can:
For seede goe and cast it, for malting not so,
   but get out the cockle, and then let it go.

Thresh barlie as yet but as neede shal require, — *Threshing of barlie.*
   fresh threshed for stoover thy cattel desire:
And therefore that threshing forbeare as ye may,
   till Candelmas comming, for sparing of hay.

Such wheat as ye keepe for the baker to buie,
   unthreshed til March in the sheafe let it lie,
Least foistnes take it if sooner yee thresh it,
   although by oft turning ye seeme to refresh it.

Save chaffe of the barlie, of wheate, and of rie, — *Chaffe of corn.*
   from feathers and foistines, where it doth lie,
Which mixed with corne, being sifted of dust,
   go give to thy cattel, when serve them ye must.

Greene peason or hastings at Hallontide sowe,
   in hartie good soile he requireth to growe:
Graie peason or runcivals cheerely to stand,
   at Candlemas sowe, with a plentifull hand.

Leave latewardly rering, keepe now no more swine,
   but such as thou maist, with the offal of thine:
Except ye have wherewith to fat them away,
   the fewer thou keepest, keepe better yee may.

To rere up much pultrie, and want the barne doore,
   is naught for the pulter and woorse for the poore.
So, now to keepe hogs and to sterve them for meate,
   is as to keepe dogs for to bawle in the streate.

As cat a good mouser is needfull in house,
   because for hir commons she killeth the mouse,
So ravening curres, as a meany doo keepe,
   makes master want meat, and his dog to kill sheepe.

Martilmas beefe.

(For Easter) at Martilmas hang up a beefe,
   for stalfed and pease fed plaie pickpurse the theefe:
With that and the like, er an grasse biefe come in,
   thy folke shal looke cheerelie when others looke thin.

Set garlike and beanes.

Set garlike and beanes, at S. Edmond the king,
   the moone in the wane, thereon hangeth a thing:
Thencrease of a pottle (wel prooved of some)
   shal pleasure thy household er peskod time come.

Threshing.

When raine is a let to thy dooings abrode,
   set threshers a threshing to laie on good lode:
Thresh cleane ye must bid them, though lesser they yarn,
   and looking to thrive, have an eie to thy barne.

Cattle beaters.

Take heede to thy man in his furie and heate, (neate:
   with ploughstaff and whipstock, for maiming thy
To thresher for hurting of cow with his flaile,
   or making thy hen to plaie tapple up taile.

Some pilfering thresher will walke with a staffe,   *Corne*
  will carrie home corne as it is in the chaffe,   *stealers.*
And some in his bottle of leather so great
  will carry home daily both barlie and wheat.

If houseroome will serve thee, lay stover up drie,   *Keepe dry*
  and everie sort by it selfe for to lie.   *thy straw.*
Or stack it for litter, if roome be too poore,
  and thatch out the residue noieng thy doore.

Cause weekly thy thresher to make up his flower,   *Everie*
  though slothfull and pilferer thereat doo lower: *weeke rid*
Take tub for a season, take sack for a shift,   *thy barne*
  yet garner for graine is the better for thrift.   *flower.*

All maner of strawe that is scattered in yard,
  good husbandlie husbands have daily regard,
In pit full of water the same to bestowe,
  where lieng to rot, thereof profit may growe.

Now plough up thy hedlond, or delve it with spade,   *Digging of*
  where otherwise profit but little is made: *hedlonds.*
And cast it up high, upon hillocks to stand,
  that winter may rot it, to compas thy land.

If garden requier it, now trench it ye may,   *Trenching*
  one trench not a yard from another go lay: *of garden.*
Which being well filled with muck by and by,
  go cover with mould for a season to ly.

Foule privies are now to be clensed and fide,   *Clensing*
  let night be appointed such baggage to hide: *of privies.*
Which buried in garden, in trenches alowe,
  shall make very many things better to growe.

The chimney all sootie would now be made cleene,   *Sootie*
  for feare of mischances, too oftentimes seene: *chimneyes.*
Old chimney and sootie, if fier once take,
  by burning and breaking, soone mischeefe may make.

*Put horse
into stable.*

> When ploughing is ended, and pasture not great,
>   then stable thy horses, and tend them with meat:
> Let season be drie when ye take them to house,
>   for danger of nittes, or for feare of a louse.

*Saving
of doong.*

> Lay compas up handsomly, round on a hill,
>   to walke in thy yard at thy pleasure and will,
> More compas it maketh and handsom the plot,
>   if horsekeeper daily forgetteth it not.

> Make hillocks of molehils, in field thorough out,
>   and so to remaine, till the yeere go about.
> Make also the like whereas plots be too hie,
>   all winter a rotting for compas to lie.

Thus endeth Novembers husbandrie.

DECEMBERS ABSTRACT

Chapter 20

NO season to hedge,
get beetle and wedge.
Cleave logs now all,
for kitchen and hall.

Dull working tooles
soone courage cooles.

Leave off tittle tattle,
and looke to thy cattle.
Serve yoong poore elves
alone by themselves.

Warme barth for neate,
woorth halfe their meate.
The elder that nurteth
the yonger soone hurteth.

Howse cow that is old,
while winter doth hold.

Out once in a day,
to drinke and to play.

Get trustie to serve,
least cattle doo sterve.
And such as in deede
may helpe at a neede.

Observe this law,
in serving out straw.

In walking about,
good forke spie out.

At full and at change,
spring tides are strange.
If doubt ye fray,
drive cattle away.

Dank ling forgot
will quickly rot.

Here learne and trie
to turne it and drie.

Now stocks remoove,
that Orchards loove.

Set stock to growe
too thick nor too lowe.
Set now, as they com,
both cherie and plom.

Sheepe, hog, and ill beast,
bids stock to ill feast.

At Christmas is good
to let thy horse blood.

Mark here what rable
of evils in stable.

Mixe well (old gaffe)
horse corne with chaffe.
Let Jack nor Gill
fetch corne at will.

Some countries gift
to make hard shift.

Some cattle well fare
with fitches and tare.
Fitches and tares
be Norfolke wares.

Tares threshed with skill
bestowe as yee will.

Hide strawberies, wife,
to save their life.

Knot, border, and all,
now cover ye shall.

Helpe bees, sweet conie,
with licour and honie.

Get campers a ball,
to campe therewithall.

*Thus endeth Decembers
abstract, agreeing with
Decembers husbandrie.*

OTHER SHORT
REMEMBRANCES

Let Christmas spie
yard cleane to lie.
No labour, no sweate,
go labour for heate.
Feede dooves, but kill not,
if stroy them ye will not.
Fat hog or (er ye kill it)
or else ye doo spill it.

Put oxe in stall,
ere oxe doo fall.
Who seeteth hir graines,
hath profit for paines.
Rid garden of mallow,
plant willow and sallow.

Let bore life render,
see brawne sod tender,
For wife, fruit bie,
for Christmas pie.

Ill bread and ill drinke,
makes many ill thinke.
Both meate and cost
ill dressed halfe lost.

Who hath wherewithall,
may cheere when he shall:
But charged man,
must cheere as he can.

Here ends Decembers short remembrances.

DECEMBERS HUSBANDRIE

Chapter 21

O dirtie December　　Forgotten month past,
For Christmas remember.　Doe now at the last.

WHEN frost will not suffer to dike and to hedge,   *Beetle and*
  then get thee a heat with thy beetle and wedge: *wedges.*
Once Hallomas come, and a fire in the hall,
  such slivers doo well for to lie by the wall.

Get grindstone and whetstone, for toole that is dull,   *Grinding*
  or often be letted and freat bellie full.   *stone and*
A wheele barrow also be readie to have   *whetston.*
  at hand of thy servant, thy compas to save.

Give cattle their fodder in plot drie and warme,   *Serving of*
  and count them for miring or other like harme.   *cattle.*
Yoong colts with thy wennels together go serve,
  least lurched by others they happen to sterve.

The rack is commended for saving of doong,   *Woodland*
  so set as the old cannot mischiefe the yoong: *countrie.*
In tempest (the wind being northly or east)
  warme barth under hedge is a sucker to beast.

The housing of cattel while winter doth hold,   *Housing*
  is good for all such as are feeble and old: *of cattel.*
It saveth much compas, and many a sleepe,   *Champion.*
  and spareth the pasture for walke of thy sheepe.

For charges so little much quiet is won,   *Champion.*
  if strongly and handsomly al thing be don:
But use to untackle them once in a day,
  to rub and to lick them, to drink and to play.

Ordering of cattel.

    Get trustie to tend them, not lubberlie squire,
        that all the day long hath his nose at the fire.
    Nor trust unto children poore cattel to feede,
        but such as be able to helpe at a neede.

    Serve riestraw out first, then wheatstraw and pease,
        then otestraw and barlie, then hay if ye please:
    But serve them with hay while the straw stover last,
        then love they no straw, they had rather to fast.

Forkes and yokes.

    Yokes, forks, and such other, let bailie spie out,
        and gather the same as he walketh about.
    And after at leasure let this be his hier,
        to beath them and trim them at home by the fier.

Going of cattel in marshes.

    As well at the full of the moone as the change,
        sea rages in winter be sodainly strange.
    Then looke to thy marshes, if doubt be to fray,
        for feare of (*ne forte*) have cattel away.

Looke to thy ling and saltfish.

    Both saltfish and lingfish (if any ye have)
        through shifting and drieng from rotting go save:
    Least winter with moistnes doo make it relent,
        and put it in hazard before it be spent.

How to use ling and haberden.

    Broome fagot is best to drie haberden on,
        lay boord upon ladder if fagots be gon.
    For breaking (in turning) have verie good eie,
        and blame not the wind, so the weather be drie.

    Good fruit and good plentie doth well in the loft,
        then make thee an orchard and cherish it oft:

Remooving of trees.

    For plant or for stock laie aforehand to cast,
        but set or remoove it er Christmas be past.

An orchard point.

    Set one fro other full fortie foote wide,
        to stand as he stood is a part of his pride.
    More faier, more woorthie, of cost to remoove,
        more steadie ye set it, more likely to proove.

To teach and unteach in a schoole is unmeete,
  to doe and undoe to the purse is unsweete.   *Orchard*
Then orchard or hopyard, so trimmed with cost,   *and hop-*
  should not through follie be spoiled and lost.   *yard.*

Er Christmas be passed let horse be let blood,   *Letting*
  for many a purpose it doth them much good.   *horse blood.*
The daie of S. Stephen old fathers did use:
  if that doe mislike thee some other daie chuse.

Looke wel to thy horses in stable thou must,   *Breeding of*
  that haie be not foistie, nor chaffe ful of dust: *the bots.*
Nor stone in their provender, feather, nor clots,
  nor fed with greene peason, for breeding of bots.

Some horsekeeper lasheth out provender so,   *Hog and*
  some Gillian spendal so often doth go.   *hennes*
For hogs meat and hens meat, for that and for this, *meate.*
  that corne loft is empted er chapman hath his.

Some countries are pinched of medow for hay,
  yet ease it with fitchis as well as they may.
Which inned and threshed and husbandlie dight,
  keepes laboring cattle in verie good plight.

In threshing out fitchis one point I will shew,
  first thresh out for seede of the fitchis a few:
Thresh few fro thy plowhorse, thresh cleane for the cow,
  this order in Norfolke good husbands alow.

If frost doe continue, take this for a lawe,   *Strawberies.*
  the strawberies looke to be covered with strawe.
Laid overly trim upon crotchis and bows,
  and after uncovered as weather allows.

The gilleflower also, the skilful doe knowe,   *Gille flowers.*
  doe looke to be covered, in frost and in snowe.
The knot, and the border, and rosemarie gaie,
  do crave the like succour for dieng awaie.

How to preserve bees.   Go looke to thy bees, if the hive be too light,
    set water and honie, with rosemarie dight.
Which set in a dish ful of sticks in the hive,
    from danger of famine yee save them alive.

In medow or pasture (to growe the more fine)
    let campers be camping in any of thine:
Which if ye doe suffer when lowe is the spring,
    you gaine to your selfe a commodious thing.

Thus endeth Decembers husbandrie.

A DIGRESSION TO HOSPITALITIE

Chapter 22

LEAVE husbandrie sleeping a while ye must doo,
    to learne of housekeeping a lesson or twoo.
What ever is sent thee by travell and paine,
    a time there is lent thee to rendrit againe.
Although ye defend it, unspent for to bee,
    another shall spend it, no thanke unto thee.
How ever we clime, to accomplish the mind,
    we have but a time thereof profit to find.

A DESCRIPTION OF TIME, AND THE YEARE

Chapter 23

OF God to thy dooings a time there is sent,
   which endeth with time that in dooing is spent.
For time is it selfe but a time for a time,
   forgotten ful soone, as the tune of a chime.

In Spring time we reare, we doo sowe, and we plant,   *Spring.*
   in Sommer get vittels, least after we want.   *Sommer.*
In Harvest we carie in corne and the fruit,   *Harvest.*
   in Winter to spend as we neede of ech suit.   *Winter.*

The yeere I compare, as I find for a truth,   *Childhood.*
   the Spring unto childhood, the Sommer to youth,   *Youth.*
The Harvest to manhood, the Winter to age: *Manhood.*
   all quickly forgot as a play on a stage.   *Age.*

Time past is forgotten, er men be aware,
   time present is thought on with woonderfull care,
Time comming is feared, and therefore we save,
   yet oft er it come, we be gone to the grave.

A DESCRIPTION OF LIFE AND RICHES

Chapter 24.

WHO living but daily discerne it he may,
   how life as a shadow doth vanish away;
And nothing to count on so suer to trust
   as suer of death and to turne into dust.

The lands and the riches that here we possesse
   be none of our owne, if a God we professe,
But lent us of him, as his talent of gold,
   which being demanded, who can it withhold?

God maketh no writing that justly doth say
   how long we shall have it, a yeere or a day;
Atrop, or death.   But leave it we must (how soever we leeve)
   when Atrop shall pluck us from hence by the sleeve.

To death we must stoupe, be we high, be we lowe,
   but how and how sodenly, few be that knowe:
What carie we then, but a sheete to the grave,
   to cover this carkas, of all that we have?

A DESCRIPTION OF HOUSEKEEPING

Chapter 25

WHAT then of this talent, while here we remaine,
   to studie to yeeld it to God with a gaine?
And that shall we doo, if we doo it not hid,
   but use and bestow it, as Christ doth us bid.

What good to get riches by breaking of sleepe,
   but (having the same) a good house for to keepe?
Not onely to bring a good fame to thy doore,
   but also the praier to win of the poore.

Of all other dooings house keeping is cheefe,
   for daily it helpeth the poore with releefe;
The neighbour, the stranger, and all that have neede,
   which causeth thy dooings the better to speede.

Though harken to this we should ever among,
   yet cheefly at Christmas, of all the yeare long.
Good cause of that use may appeare by the name,
   though niggerly niggards doo kick at the same.

A DESCRIPTION
of the feast of the birth of Christ,
commonly called Christmas.

Chapter 26

OF Christ commeth Christmas, the name with the feast,
   a time full of joie to the greatest and least:
At Christmas was Christ (our Saviour) borne,
   the world through sinne altogether forlorne.

At Christmas the daies doo begin to take length,
   of Christ doth religion cheefly take strength.
As Christmas is onely a figure or trope,
   so onely in Christ is the strength of our hope.

At Christmas we banket, the rich with the poore,
   who then (but the miser) but openeth [h]is doore?
At Christmas of Christ many Carols we sing,
   and give many gifts in the joy of that King.

At Christmas in Christ we rejoice and be glad,
   as onely of whom our comfort is had;
At Christmas we joy altogether with mirth,
   for his sake that joyed us all with his birth.

A DESCRIPTION OF APT TIME TO SPEND

Chapter 27

LET such (so fantasticall) liking not this,
  nor any thing honest that ancient is,
Give place to the time that so meete we doo see
  appointed of God as it seemeth to bee.

At Christmas good husbands have corne on the ground,
  in barne, and in soller, woorth many a pound,
With plentie of other things, cattle and sheepe,
  all sent them (no doubt on) good houses to keepe.

At Christmas the hardnes of Winter doth rage,
  a griper of all things and specially age:
Then lightly poore people, the yoong with the old,
  be sorest oppressed with hunger and cold.

At Christmas by labour is little to get,
  that wanting, the poorest in danger are set.
What season then better, of all the whole yeere,
  thy needie poore neighbour to comfort and cheere?

AGAINST FANTASTICALL SCRUPLENES

Chapter 28

AT this time and that time some make a great matter,
   som help not but hinder the poore with their clatter.
Take custome from feasting, what commeth then last,
   where one hath a dinner, a hundred shall fast.

To dog in the manger some liken I could,
   that hay will eate none, nor let other that would;
Some scarce in a yeere give a dinner or twoo,
   nor well can abide any other to doo.

Play thou the good fellow, seeke none to misdeeme,
   disdaine not the honest, though merie they seeme:
For oftentimes seene, no more verie a knave
   than he that doth counterfait most to be grave.

CHRISTMAS HUSBANDLIE FARE

Chapter 29

GOOD husband and huswife now cheefly be glad,
   things handsom to have, as they ought to be had;
They both doo provide against Christmas doo come,
   to welcome good neighbour, good cheere to have some.

Good bread and good drinke, a good fier in the hall,   *Christmas*
   brawne, pudding and souse, and good mustard withall.   *cuntrie fare.*

Beefe, mutton, and porke, shred pies of the best,
   pig, veale, goose and capon, and turkey well drest;
Cheese, apples and nuts, joly Carols to heare,
   as then in the countrie is counted good cheare.

What cost to good husband is any of this?
   good houshold provision onely it is.
Of other the like, I doo leave out a menie,
   that costeth the husbandman never a penie.

A CHRISTMAS CAROLL
of the birth of Christ
upon the tune of King Salomon

Chapter 30

WAS not Christ our Saviour
sent to us fro God above?
not for our good behaviour,
but onely of his mercie and love.
If this be true, as true it is,
    truely in deede,
great thanks to God to yeeld for this,
    then had we neede.

This did our God for very troth,
to traine to him the soule of man,
and justly to performe his oth
to Sara and to Abram than,
That through his seed all nations should
    most blessed bee:
As in due time peforme he would
    as now wee see.

Which woonderously is brought to pas,
and in our sight alredie donne,
by sending as his promise was
(to comfort us) his onely sonne,
Even Christ (I meane) that virgins child,
    in Bethlem borne,
that Lambe of God, that Prophet mild,
    with crowned thorne.

Such was his love to save us all,
from dangers of the curse of God,
that we stood in by Adams fall,
and by our owne deserved rod,
That through his blood and holie name
    who so beleeves,
and flie from sinne and abhors the same,
    free mercie he geeves.

For these glad newes this feast doth bring:
to God the Sonne and holy Ghost
let man give thanks, rejoice, and sing,
from world to world, from cost to cost:
for all good gifts so many waies
    that God doth send,
let us in Christ give God the praies,
    till life shall end.

T. Tusser.

At Christmas be merie and thankfull withall,
And feast thy poore neighbors, the great with the small,
Yea, all the yeere long, to the poore let us give,
Gods blessing to folow us while wee doo live.

JANUARIES ABSTRACT

Chapter 31

BID Christmas adew,
thy stock now renew.

Who killeth a neat,
hath cheaper his meat.
Fat home fed souse,
is good in a house.

Who dainties love,
a begger shall prove.
Who alway selles,
in hunger dwelles.

Who nothing save,
shall nothing have.

Lay durt upon heapes,
some profit it reapes.
When weather is hard,
get muck out of yard.
A fallow bestowe,
where pease shall growe.
Good peason and white,
a fallow will quite.

Go gather quickset,
the yongest go get.
Dig garden, stroy mallow,
set willow and sallow.
Greene willow for stake
in bank will take.

Let Doe go to buck,
with Conie good luck.
Spare labour nor monie,
store borough with conie.
Get warrener bound
to vermin thy ground.
Feed Doves, but kill not,
if loose them ye will not.
Dove house repaire,
make Dovehole faire.
For hop ground cold,
Dove doong woorth gold.

Good gardiner mine,
make garden fine.
Set garden pease,
and beanes if ye please.
Set Respis and Rose,
yoong rootes of those.

The timelie buier
hath cheaper his fier.

Some burns without wit,
some fierles sit.

Now season is good
to lop or fell wood.
Prune trees some allows
for cattle to brows.

Give sheepe to their fees
the mistle of trees.

Let lop be shorne
that hindreth corne.
Save edder and stake,
strong hedge to make.

For sap as ye knowe,
let one bough growe.
Next yeere ye may
that bough cut away.

A lesson good
to encrease more wood.

Save crotchis of wud,
save spars and stud.
Save hop for his dole,
the strong long pole.

How ever ye scotch,
save pole and crotch.

From Christmas to May,
weake cattle decay.

With vergis acquaint
poore bullock so faint;
This medcin approved
is for to be looved.

Let plaister lie
three daies to trie:
too long if ye stay,
taile rots away.

Eawes readie to yeane
craves ground rid cleane.
Keepe sheepe out of briers,
Keepe beast out of miers.

Keepe bushes from bill,
till hedge ye will:
Best had for thy turne,
their rootes go and burne.

No bushes of mine,
if fence be thine.

In stubbed plot,
fill hole with clot.

Rid grasse of bones,
of sticks and stones.

Warme barth give lams,
good food to their dams,
Look daily well to them,
least dogs undoo them.

Yoong lamb well sold,
fat lamb woorth goold.

Keepe twinnes for breed,
as eawes have need.

One calfe if it please ye,
now reared shall ease ye.
Calves likely reare,
at rising of yeare.
Calfe large and leane
is best to weane.

Calfe lickt take away,
and howse it ye may.
This point I allow
for servant and cow.

Calves yonger than other
learne one of another.

No danger at all
to geld as they fall.
Yet Michel cries
please butchers eies.

Sow ready to fare,
craves huswives care.

Leave sow but five,
the better to thrive.

Weane such for store
as sucks before.
Weane onely but three
large breeders to bee.

Lamb, bulchin, and pig,
geld under the big.

Learne wit, sir dolt,
in gelding of colt.

Geld yoong thy filly,
else perish will ginny.
Let gelding alone,
so large of bone.
By breathely tits
few profit hits.

Breede ever the best,
and doo of the rest,
Of long and large,
take huswife a charge.

Good cow & good ground
yeelds yeerely a pound.
Good faring sow
holds profit with cow.

Who keepes but twaine,
the more may gaine.

Tith justly, good garson,
else drive will the parson.

Thy garden twifallow,
stroy hemlock and mallow.

Like practise they proove,
that hops doe loove.

Now make and wand in
trim bower to stand in.
Leave wadling about,
till arbor be out.

Who now sowes otes,
gets gold and grotes.
Who sowes in May
gets little that way.

Go breake up land,
get mattock in hand,
Stub roote so tough,
for breaking of plough.

What greater crime
then losse of time?

Lay land or lease
breake up if ye please.
But fallow not yet,
that hast any wit.

Where drink ye sowe,
good tilth bestowe.

Small profit is found,
by peeling of ground.

Land past the best
cast up to rest.

*Thus endeth Januaries
abstract, agreeing with
Januaries husbandrie.*

OTHER SHORT
REMEMBRANCES

Get pulling hooke (sirs),
for broome and firs.
Pluck broome, broome still,
cut broome, broome kill.

Broome pluckt by and by,
breake up for rie.
Friend ringle thy hog,
or looke for a dog.

In casting provide,
for seede lay aside.
Get doong, friend mine,
for stock and vine.

If earth be not soft,
go dig it aloft.
For quamier get bootes,
stub alders and rootes.

Hop poles waxe scant,
for poles mo plant.
Set chestnut and walnut,
set filbeard and smalnut.

Peach, plumtree, & cherie,
yoong bay and his berie.
Or set their stone,
unset leave out none.

Sowe kirnels to beare,
of apple and peare.
All trees that beare goom
set now as they coom.

Now set or remoove
such stocks as ye loove.

*Here ends Januaries short
remembrances.*

OF TREES OR FRUITES TO BE SET OR REMOVED

Apple trees of all sorts.
Apricocks.
Barberies.
Boollesse, black & white.
Cheries, red and black.
Chestnuts.
Cornet plums.
Damsens, white & black.
Filbeards, red and white.
Goose beries.
Grapes, white and red.
Greene or grasse plums.
Hurtillberies.
Medlars or marles.

Mulberie.
Peaches, white and red.
Peares of all sorts.
Perareplums, black & yelow.
Quince trees.
Respis.
Reisons.
Small nuts.
Strawberies, red and white.
Service trees.
Walnuts.
Wardens, white and red.
Wheat plums.

Now set ye may
the box and bay,
Haithorne and prim,
for clothes trim.

JANUARIES HUSBANDRIE

Chapter 32

A kindly good Janiveere,   *Forgotten month past.*
Freeseth pot by the feere.   *Doe now at the last.*

WHEN Christmas is ended, bid feasting adue,   *Husbandly*
  goe play the good husband, thy stock to renue.   *lessons.*
Be mindfull of rearing, in hope of a gaine,
  dame profit shall give thee reward for thy paine.

Who both by his calfe and his lamb will be knowne,
  may well kill a neate and a sheepe of his owne.
And he that can reare up a pig in his house,
  hath cheaper his bacon and sweeter his souse.

Who eateth his veale, pig and lamb being froth,
  shall twise in a weeke go to bed without broth.
Unskilfull that passe not, but sell away sell,
  shall never have plentie where ever they dwell.

Be greedie in spending, and careles to save,
  and shortly be needie and readie to crave.
Be wilfull to kill and unskilfull to store,
  and looke for no foison, I tell thee before.

Lay dirt upon heapes, faire yard to be seene,
  if frost will abide it, to feeld with it cleene.
In winter a fallow some love to bestowe,
  where pease for the pot they intend for to sowe.

In making or mending as needeth thy ditch,   *Quick set*
  get set to quick set it, learne cunningly whitch.   *now.*
In hedging (where clay is) get stake as ye knowe,
  of popler and willow, for fewell to growe.

Keepe cleane thy dovehous.	Leave killing of conie, let Doe go to buck,
	   and vermine thy burrow, for feare of ill luck.
	Feed Dove (no more killing), old Dove house repaire,
	   save dove dong for hopyard, when housey emake faire.

Runcival peason.	Dig garden, stroy mallow, now may ye at ease,
	   and set (as a daintie) thy runcivall pease.
	Go cut and set roses, choose aptly thy plot,
	   the rootes of the yoongest are best to be got.

Timelie provision for fewell.	In time go and bargaine, least woorser doo fall,
	   for fewell, for making, for carriage and all.
	To buie at the stub is the best for the buier,
	   more timelie provision, the cheaper is fier.

Ill husbandrie.	Some burneth a lode at a time in his hall,
	   some never leave burning til burnt they have all.
	Some making of havock, without any wit,
	   make many poore soules without fire to sit.

Pruning of trees.	If frost doo continue, this lesson doth well,
	   for comfort of cattel the fewell to fell:
	From everie tree the superfluous bows
	   now prune for thy neat thereupon to go brows.

	In pruning and trimming all maner of trees, reserve to ech cattel their properly fees.
Mistle and Ivie.	If snowe doo continue, sheepe hardly that fare crave Mistle and Ivie for them for to spare.

Lopping of pollengers.	Now lop for thy fewell old pollenger growen,
	   that hinder the corne or the grasse to be mowen.
	In lopping and felling, save edder and stake,
	   thine hedges as needeth to mend or to make.

	In lopping, old Jocham, for feare of mishap,
	   one bough stay unlopped, to cherish the sap:
	The second yeere after then boldly ye may,
	   for driping his fellowes, that bough cut away.

Lop popler and sallow, elme, maple, and prie,   *The propertie*
  well saved from cattle, till Sommer to lie.   *of soft wood.*
So far as in lopping, their tops ye doo fling,
  so far without planting yoong copie will spring.

Such fewell as standing a late ye have bought,
  now fell it, and make it, and doo as ye ought.
Give charge to the hewers (that many things mars),
  to hew out for crotches, for poles, and for spars.

If hopyard or orchard ye mind for to have,   *Hoppoles*
  for hoppoles and crotches in lopping go save.   *and*
Which husbandlie spared may serve at a push,   *crotches.*
  and stop by so having two gaps with a bush.

From Christmas, till May be well entered in,
  some cattle waxe faint, and looke poorely and thin.
And cheefly when prime grasse at first doth appeere,
  then most is the danger of all the whole yeere.

Take vergis and heate it, a pint for a cow,   *A medicen*
  bay salt a hand full, to rub tong ye wot how.   *for faint*
That done, with the salt, let hir drinke off the rest: *cattell.*
  this manie times raiseth the feeble up best.

Poore bullock with browsing and naughtily fed,   *To fasten*
  scarce feedeth, hir teeth be so loose in hir hed: *loose teeth*
Then slise ye the taile where ye feele it so soft,   *in a bullock.*
  with soote and with garlike bound to it aloft.

By brembles and bushes, in pasture too full,   *Ewes upon*
  poore sheepe be in danger and loseth their wull.   *eaning.*
Now therefore thine ewe, upon lamming so neere,
  desireth in pasture that all may be cleere.

Leave grubbing or pulling of bushes (my sonne)
  till timely thy fences require to be donne.
Then take of the best, for to furnish thy turne,
  and home with the rest, for the fier to burne.

	In everie greene, if the fence be not thine,
Stubbing of greenes.	now stub up the bushes, the grasse to be fine.

Least neighbour doo dailie so hack them believe,
    that neither thy bushes nor pasture can thrive.

In ridding of pasture with turfes that lie by,
    fill everie hole up, as close as a dy.
The labour is little, the profit is gay,
    what ever the loitering labourers say.

The sticks and the stones go and gather up cleene,
    for hurting of sieth or for harming of greene.
For feare of Hew prowler, get home with the rest,
    when frost is at hardest, then carriage is best.

Yoong lambes.

Yoong broome or good pasture thy ewes doo require,
    warme barth and in safetie their lambes doo desire.
Looke often well to them, for foxes and dogs,
    for pits and for brembles, for vermin and hogs.

More daintie the lambe, the more woorth to be sold,
    the sooner the better for eaw that is old.
But if ye doo minde to have milke of the dame,
    till Maie doo not sever the lambe fro the same.

Rearing of lambs.

Ewes yeerly by twinning rich maisters doo make,
    the lamb of such twinners for breeders go take.
For twinlings be twiggers, encrease for to bring,
    though som for their twigging *Peccantem* may sing.

Rearing of calves.

Calves likely that come between Christmas and Lent,
    take huswife to reare, or else after repent:
Of such as doo fall betweene change and the prime,
    no rearing, but sell or go kill them in time.

Howsing of cattel.

Howse calfe, and go sockle it twise in a day,
    and after a while, set it water and hay.
Stake ragged to rub on, no such as will bend,
    then weane it well tended, at fiftie daies end.

The senior weaned his yoonger shall teach,
  how both to drinke water and hay for to reach.
More stroken and made of when ought it doo aile,
  more gentle ye make it, for yoke or the paile.

Geld bulcalfe and ramlamb, as soone as they fall, *Of gelding.*
  for therein is lightly no danger at all.
Some spareth the ton for to pleasure the eie,
  to have him shew greater when butcher shall bie.

Sowes readie to farrow this time of the yeere
  are for to be made of and counted full deere.
For now is the losse of a fare of the sow
  more great then the losse of two calves of thy cow.

Of one sow togither reare few above five, *Rearing of*
  and those of the fairest and likest to thrive. *pigs.*
Ungelt of the best keepe a couple for store,
  one bore pig and sow pig, that sucketh before.

Who hath a desire to have store verie large, *A way to*
  at Whitsontide let him give huswife a charge, *have large*
To reare of a sow at once onely but three, *breed of*
  and one of them also a bore let it bee. *hogs.*

Geld under the dam, within fortnight at least, *Gelding*
  and save both thy monie and life of the beast. *time.*
Geld later with gelders as many one do,
  and looke of a doozen to geld away two.

Thy colts for thy saddle geld yoong to be light, *Gelding of*
  for cart doo not so, if thou judgest aright. *horse coltes.*
Nor geld not but when they be lustie and fat:
  for there is a point, to be learned in that.

Geld fillies (but tits) er an nine daies of age, *Gelding of*
  they die else of gelding (or gelders doo rage). *fillies.*
Yoong fils so likelie of bulke and of bone:
  keepe such to be breeders, let gelding alone.

Reare the fairest of al things.

For gaining a trifle, sell never thy store,
    what joy to acquaintance, what pleasureth more?
The larger of bodie, the better for breede:
    more forward of growing, the better they speede.

Of cow and sow.

Good milchcow, well fed, that is faire and sound,
    is yeerely for profit as good as a pound:
And yet by the yeere, I have prooved er now,
    as good to the purse is a sow as a cow.

Keepe one and keepe both, with as little a cost,
    then all shall be saved and nothing be lost.
Both having togither what profit is caught,
    good huswifes (I warrant ye) need not be taught.

For lamb, pig and calfe, and for other the like,
    tithe so as thy cattle the Lord doo not strike.
Or if yee deale guilefully, parson will dreve,
    and so to your selfe a worse turne ye may geve.

Thy garden plot latelie well trenched and muckt,
    would now be twifallowd, the mallowes out pluckt,
Well clensed and purged of roote and of stone,
    that falt therein afterward found may be none.

Weeding of hopyard.

Remember thy hopyard, if season be drie,
    now dig it and weed it, and so let it lie.
More fennie the laier the better his lust,
    more apt to beare hops when it crumbles like dust.

Trimming up arbors.

To arbor begun, and quick setted about,
    no poling nor wadling till set be far out.
For rotten and aged may stand for a shew,
    but hold to their tackling there doe but a few.

Sowing of otes. Late sowing not good.

In Janivere husband that poucheth the grotes
    will break up his laie, or be sowing of otes,
Otes sowen in Janivere, laie by the wheat,
    in May by the hay for the cattle to eat.

Let servant be readie, with mattock in hand,
   to stub out the bushes that noieth the land:
And cumbersome rootes, so annoieng the plough,
   turne upward their arses with sorrow inough.

Who breaketh up timelie his fallow or lay, *Breaking up*
   sets forward his husbandrie many a way. *lay in som*
This trimlie well ended doth forwardly bring, *countrie.*
   not onelie thy tillage, but all other thing.

Though lay land ye breke up when Christmas is gon,
   for sowing of barlie or otes thereupon,
Yet haste not to fallow til March be begun,
   least afterward wishing it had ben undun.

Such land as ye breake up for barlie to sowe,
   two earthes at the least er ye sowe it bestowe.
If land be thereafter, set oting apart,
   and follow this lesson, to comfort thine hart.

Some breaking up laie soweth otes to begin,
   to suck out the moisture so sower therein.
Yet otes with hir sucking a peeler is found,
   both ill to the maister and worse to som ground.

Land arable driven or worne to the proofe,
   and craveth some rest for thy profits behoofe.
With otes ye may sowe it, the sooner to grasse,
   more soone to be pasture to bring it to passe.

   Thus endeth Januaries husbandrie.

FEBRUARIES ABSTRACT

Chapter 33

LAY compas ynow,
er ever ye plow.

Place doong heapes alowe,
more barlie to growe.

Eat etch er ye plow,
with hog, sheepe and cow.
Sowe lintels ye may,
and peason gray.
Keepe white unsowne,
till more be knowne.

Sow pease (good trull)
the Moone past full.
Fine seedes then sowe,
whilst Moone doth growe.

Boy, follow the plough,
and harrow inough.
So harrow ye shall,
till coverd be all.

Sowe pease not too thin,
er plough ye set in.

Late sowen sore noieth,
late ripe, hog stroieth.

Some provender save,
for plowhorse to have.

To oxen that drawe,
give hay and not strawe.
To steeres ye may
mixe strawe with hay.

Much carting, ill tillage,
makes som to flie village.

Use cattle aright,
to keepe them in plight.

Good quickset bie,
old gatherd will die.

Stick bows a rowe,
where runcivals growe.

Sowe kirnels and hawe,
where ridge ye did drawe.

Sowe mustard seed,
and helpe to kill weed.
Where sets doo growe,
see nothing ye sowe.

Cut vines and osier,
plash hedge of enclosier.
Feed highly thy swan,
to love hir good man.
Nest high I advise,
least floud doe arise.

Land meadow spare,
there doong is good ware.

Go strike off the nowles
of delving mowles.
Such hillocks in vaine
lay leavelled plaine.

To wet the land,
let mowle hill stand.

Poore cattle crave
some shift to have.

Cow little giveth
that hardly liveth.

Rid barlie al now,
cleane out of thy mow.
Choice seed out drawe,
save cattle the strawe.

To coast man ride
Lent stuffe to provide.

*Thus endeth Februaries
abstract, agreeing with
Februaries husbandrie.*

OTHER SHORT
REMEMBRANCES

Trench medow and redge,
dike, quickset, and hedge.
To plots not full,
ad bremble and hull.

Let wheat and the rie
for thresher still lie.
Such strawe some save,
for thacker to have.

Poore cunnie, so bagged,
is soone over lagged.
Plash burrow, set clapper,
for dog is a snapper.

Good flight who loves,
must feed their doves.
Bid hauking adew,
cast hauke into mew.

Keepe sheepe out of briers,
keepe beast out of miers.
Keepe lambes from fox,
else shepherd go box.

Good neighbour mine,
now yoke thy swine.
Now everie day,
set hops ye may.

Now set for thy pot,
best herbes to be got.

For flowers go set,
all sorts ye can get.

As winter doth proove,
so may ye remoove.
Now all things reare,
for all the yeare.

Watch ponds, go looke
to weeles and hooke.
Knaves seld repent
to steale in Lent.

Alls fish they get
that commeth to net.
Who muck regards
makes hillocks in yards.

Here ends Februaries short remembrances.

FEBRUARIES HUSBANDRIE

Chapter 34

Feb, fill the dike   *Forgotten month past,*
With what thou dost like.   *Doe now at the last.*

WHO laieth on doong er he laieth on plow,
   such husbandrie useth as thrift doth alow.
One month er ye spred it, so still let it stand,
   er ever to plow it, ye take it in hand.

Place doong heape a low by the furrough along.
   where water all winter time did it such wrong.
So make ye the land to be lustie and fat,
   and corne thereon sowen to be better for that,

Go plow in the stubble, for now is the season,
   for sowing of fitchis, of beanes, and of peason.
Sowe runcivals timelie, and all that be gray,
   but sowe not the white till S. Gregories day.

Sowe peason and beanes in the wane of the Moone,
   who soweth them sooner, he soweth too soone.
That they with the planet may rest and arise,
   and flourish with bearing most plentifull wise.

Friend, harrow in time, by some maner of meanes,
   not onely thy peason, but also thy beanes.
Unharrowed die, being buried in clay,
   where harrowed florish, as flowers in May.

Both peason and beanes sowe afore ye doo plow,
   the sooner ye harrow, the better for yow.
White peas on so good for the purse and the pot:
   let them be well used else well doo ye not.

Have eie unto harvest what ever ye sowe,
   for feare of mischances, by riping too slowe.
Least corne be destroied, contrarie to right,
   by hogs or by cattel, by day or by night.

Good provender labouring horses would have,
   good haie and good plentie, plow oxen doo crave.
To hale out the muck and to plow up thy ground :
   or else it may hinder thee many a pound.

Who slacketh his tillage, a carter to bee,
   for grote got abrode, at home lose shall three.
And so by his dooing he brings out of hart
   both land for the corne and horse for the cart.

Who abuseth his cattle and sterves them for meat,
   by carting or plowing, his gaine is not great.
Where he that with labour can use them aright,
   hath gaine to his comfort, and cattle in plight.

Buie quickset at market, new gatherd and small,
   buie bushes or willow, to fence it withall.
Set willowes to growe, in the steede of a stake,
   for cattel in sommer, a shadow to make.

Runcival peason.

Stick plentie of bows among runcivall pease
   to climber thereon, and to branch at their ease.
So dooing, more tender and greater they wex,
   if peacock and turkey leave jobbing their bex.

Now sowe and go harrow (where redge ye did draw)
   the seed of the bremble, with kernell and haw.
Which covered overlie, soone to shut out,
   goe see it be ditched and fenced about.

Sowe mustard seede.

Where banks be amended and newly up cast,
   sow mustard seed, after a shower be past.
Where plots full of nettles be noisome to eie,
   sowe thereupon hempseed, and nettle will die.

The vines and the osiers cut and go set,   *Cut or set*
  if grape be unpleasant, a better go get.   *vines.*
Feed swan, and go make hir up strongly a nest,
  for feare of a floud, good and high is the best.

Land meadow that yeerly is spared for hay,   *Catching*
  now fence it and spare it, and doong it ye may.   *of mowls.*
Get mowle catcher cunninglie mowle for to kill,
  and harrow and cast abrode everie hill.

Where meadow or pasture to mowe ye doo laie,
  let mowle be dispatched some maner of waie.
Then cast abrode mowlhill, as flat as ye can,
  for many commodities following than.

If pasture by nature is given to be wet,
  then bare with the mowlhill, though thick it be set.
That lambe may sit on it, and so to sit drie,
  or else to lie by it, the warmer to lie.

Friend, alway let this be a part of thy care,   *Looke well*
  for shift of good pasture, lay pasture to spare.   *to thy fence.*
So have you good feeding, in bushets and lease,
  and quickly safe finding of cattel at ease.

Where cattel may run about, roving at wil,
  from pasture to pasture, poor bellie to fil,
There pasture and cattel both hungrie and bare,
  for want of good husbandrie worser doo fare.

Now thresh out thy barlie, for malt or for seed,
  for bread corne (if need be) to serve as shall need.
If worke for the thresher ye mind for to have,
  of wheat and of mestlen unthreshed go save.

Now timelie for Lent stuffe thy monie disburse,
  the longer ye tarie for profit the wurse,
If one penie vantage be therein to save,
  of coast man or fleming be sure to have.

Thus endeth Februaries husbandrie.

MARCHES ABSTRACT

Chapter 35

WHITE peason sowe,
scare hungry crow.

Spare meadow for hay,
spare marshes at May.

Keepe sheepe from dog,
keepe lambes from hog.
If foxes mowse them,
then watch or howse them.

March drie or wet,
hop ground go set.
Yoong rootes well drest
proove ever best.
Grant hop great hill
to growe at will.
From hop long gut
away go cut.

Here learne the way
hop rootes to lay.

Rootes best to proove,
thus set I loove.

Leave space and roome,
to hillock to coome.

Of hedge and willow
hop makes his pillow.

Good bearing hop
climes up to the top.
Keepe hop from sunne,
and hop is undunne.

Hop tooles procure
that may endure.
Iron crowe like a stake,
deepe hole to make.
A scraper to pare
the earth about bare.
A hone to raise roote,
like sole of a boote.
Sharpe knife to cut
superfluous gut.

Who graffing looves,
now graffing prooves.
Of everie suite,
graffe daintie fruite.
Graffe good fruite all,
or graffe not at all.
Graffe soone may be lost,
both graffing and cost.
Learne here take heed
what counsell doth beed.

Sowe barlie that can,
too soone ye shall ban.
Let horse keepe his owne,
till barlie be sowne.

Sowe even thy land,
with plentifull hand.
Sowe over and under,
in claie is no woonder.

By sowing in wet,
is little to get.

Straight folow the plough,
and harrow inough.
With sling go throwe,
to scare away crowe.

Rowle after a deaw,
when barlie doth sheaw.
More handsom to make it,
to mowe and to rake it.

Learne here ye may
best harrowing way.

Now rowle thy wheat,
where clods be too great.

Make readie a plot,
for seeds for the pot.

Best searching minds
the best waie finds.

For garden best
is south southwest.

Good tilth brings seedes,
evill tilture, weedes.

For sommer sowe now,
for winter see how.

Learne time to knowe,
to set or sowe.

Yoong plants soone die,
that growes too drie.

In countrie doth rest,
what season is best.

Good peason and leekes
makes pottage for creekes.

Have spoone meat inough,
for cart and the plough.
Good poore mans fare,
is poore mans care.
And not to boast,
of sod and roast.

Cause rooke and raven
to seeke a new haven.

*Thus endeth Marches
abstract, agreeing with
Marches husbandrie.*

OTHER SHORT
REMEMBRANCES

Geld lambes now all,
straight as they fall.
Looke twise a day,
least lambes decay.

Where horse did harrow,
put stones in barrow,
And laie them by,
in heapes on hy.

Let oxe once fat
lose nothing of that.
Now hunt with dog,
unyoked hog.

With Doves good luck,
reare goose and duck.
To spare aright
spare March his flight.

The following additional couplets are in 1577 *edition.*

Save chikins poore buttocks
from pye, crowe, & puttocks.

Some love now best
yong rabbets nest.

Now knaves will steale
pig, lamb, and veale.

Here learne to knowe
what seedes to sowe.

And such to plant
whose seedes do want.

SEEDES AND HERBES
FOR THE KITCHEN

Avens.
Betanie.
Bleets or beets, white or yellow
Bloodwoort.
Buglas.
Burnet.
Burrage.
Cabage remove in June.
Clarie.
Coleworts.
Cresses.
Endive.
Fenell.
French Malows.
French Saffron set in August.
Langdebiefe.
Leekes remove in June.
Lettis remove in May.
Longwort.
Liverwort.
Marigolds often cut.
Mercurie.
Mints at all times.
Nep.
Onions from December to March.
Orach or arach, redde and white.
Patience.
Perceley.
Peneriall.
Primerose.
Poret.

Rosemary in the spring time to growe south or west.
Sage red and white.
English Saffron set in August.
Summer saverie.
Sorell.
Spinage.
Suckerie.
Siethes.
Tanzie.
Time.
Violets of all sorts.
Winter saverie.

HERBES AND ROOTES FOR SALLETS AND SAUCE

Alexanders, at all times.
Artichoks.
Blessed thistle, or *Carduus benedictus*.
Cucumbers in April & May.
Cresies, sowe with Lettice in the spring.
Endive.
Mustard seede, sowe in the spring and at Mihelmas.
Musk million, in April and May.
Mints.
Purslane.
Radish, & after remove them.
Rampions.
Rokat, in April.
Sage.
Sorell.
Spinage, for the sommer.
Sea holie.
Sperage, let growe two yeares, and then remove.
Skirrets, set these plants in March.
Suckerie.
Tarragon, set in slippes in March.
Violets of all coulors.

These buie with the penie,
Or looke not for anie.

Capers.
Lemmans.
Olives.
Orengis.
Rise.
Sampire.

HERBES AND ROOTES TO BOILE OR TO BUTTER

Beanes, set in winter.
Cabbegis, sowe in March, and after remoove.
Carrets.
Citrons, sowe in May.
Goordes in May.
Navewes sowe in June.
Pompions in May.
Perseneps in winter.

Runcivall pease set in winter.
Rapes sowe in June.
Turneps in March & April.

HERBES, BRANCHES, AND FLOWERS, FOR WINDOWES AND POTS

Baies, sowe or set in plants in Januarie.
Batchelers buttons.
Botles, blew, red, and tawnie.
Collembines.
Campions.
Cousleps.
Daffadondillies.

STROWING HERBES OF ALL SORTES

Bassel, fine and busht, sowe in May.
Baulme, set in March.
Camamel.
Costmarie.
Cousleps and paggles.
Daisies of all sorts.
Sweete fennell.
Garmander.
Isop, set in Februarie.
Lavender.
Lavender spike.
Lavender cotten.
Majerom knotted, sowe or set at the spring.
Mawdelin.
Penal riall.
Roses of all sorts, in Januarie and September.
Red mints.
Sage.
Tanzie.
Violets.
Winter saverie.

Eglantine, or sweet brier.
Fetherfew.
Flower armor sowe in May.
Flower de luce.
Flower gentle, white and red.
Flower nice.
Gileflowers, red, white & carnations, set in spring, and at Harvest in pots, pailes or tubs, or for sommer in beds.
Holiokes, red, white and carnations.
Indian eie, sowe in May, or set in slips in March.
Lavender of all sorts.
Larkes foot.
Laus tibi.
Lillium cum valium.
Lillies, red and white, sowe or set in March and September.
Marigolds double.
Nigella Romana.
Pauncies or hartesease.
Paggles, greene and yelow.

Pinkes of all sorts.
Queenes gilleflowers.
Rosemarie.
Roses of all sorts.
Snag dragons.
Sops in wine.
Sweete Williams.
Sweete Johns.
Star of Bethelem.
Star of Jerusalem.
Stocke gilleflowers of all sorts.
Tuft gilleflowers.
Velvet flowers,
   or french Marigolds.
Violets, yellow and white.
Wall gilleflowers of all sorts.

HERBES TO STILL IN SOMMER

Blessed thistle.
Betanie.
Dill.
Endive.
Eiebright.
Fennell.
Fumetorie.
Isop.
Mints.
Plantine.
Roses red and damaske.
Respies.
Saxefrage.
Strawberies.
Sorell.
Suckerie.
Woodrofe for sweete
   waters and cakes.

NECESSARIE HERBES TO GROWE IN THE GARDEN FOR PHYSICK, NOT REHERSED BEFORE

Annis.
Archangel.
Betanie.
Charviel.
Cinqfile.
Cummin.
Dragons.
Detanie, or garden ginger.
Gromel seed, for the stone.
Hartstong.
Horehound.
Lovage for the stone.
Licoras.
Mandrake.
Mogwort
Pionees.
Poppie.
Rew.
Rubarb.
Smalach, for swellings.
Saxefrage, for the stone.
Savin, for the bots.
Stitchwort.
Valerian.
Woodbine.

Thus ends in breefe,
Of herbes the cheefe,
To get more skill,
Read whom ye will,
Such mo to have,
Of field go crave.

MARCHES HUSBANDRIE

Chapter 36

March dust to be sold,   *Forgotten month past,*
Worth ransome of gold.   *Doe now at the last.*

WHITE peason, both good for the pot and the purse,
    by sowing too timelie, proove often the wurse.
Bicause they be tender and hateth the cold,
    proove March er ye sowe them, for being too bold.

Spare eating   Spare meadow at Gregorie, marshes at Pask,
of meadowe.   for feare of drie Sommer, no longer time ask.
                 Then hedge them and ditch them, bestow thereon
                         pence:
                        corne, meadow and pasture, aske alway good fence.

In Lent   Of mastives and mungrels, that manie we see,
have an ey   a number of thousands too manie there bee.
to sheep   Watch therefore in Lent, to thy sheepe go and looke,
biters.   for dogs will have vittles, by hooke or by crooke.

Setting of   In March at the furdest, drie season or wet,
hops.   hop rootes so well chosen, let skilfull go set.
                   The goeler and yonger the better I love;
                     well gutted and pared, the better they prove.

Some laieth them croswise, along in the ground,
    as high as the knee they doo cover up round.
Some prick up a stick in the mids of the same,
    that little round hillock the better to frame.

Some maketh a hollownes, halfe a foot deepe,
    with fower sets in it, set slant wise a steepe:
One foot from another, in order to lie,
    and thereon a hillock, as round as a pie.

Five foot from another ech hillock would stand,
  as straight as a leaveled line with the hand.
Let everie hillock be fower foot wide,
  the better to come to on everie side.

By willowes that groweth thy hopyard without,
  and also by hedges thy meadowes about.
Good hop hath a pleasure to climbe and to spred,
  if Sunne may have passage to comfort hir hed.

Get crowe made of iron, deepe hole for to make, *Hop tools.*
  with crosse overthwart it, as sharpe as a stake.
A hone and a parer, like sole of a boote,
  to pare away grasse and to raise up the roote.

In March is good graffing, the skilfull doo knowe, *Graffing.*
  so long as the wind in the East doo not blowe.
From Moone being changed til past be the prime,
  for graffing and cropping is verie good time.

Things graffed or planted, the greatest and least,
  defend against tempest, the bird and the beast.
Defended shall prosper, the tother is lost,
  the thing with the labour, the time and the cost.

Sowe barlie in March, in April and Maie, *Sowing of*
  the latter in sand, and the sooner in claie. *barlie.*
What worser for barlie than wetnes and cold?
  what better to skilfull than time to be bold?

Who soweth his barlie too soone or in raine,
  of otes and of thistles shall after complaine.
I speake not of Maie weed, cockle and such,
  that noieth the barlie, so often and much.

Let barlie be harrowed, finelie as dust,
  then workmanly trench it and fence it ye must.
This season well plied, set sowing an end,
  and praise and praie God a good harvest to send.

Rowling of barlie.   Some rowleth their barlie straight after a raine,
    when first it appeareth to leavell it plaine.
The barlie so used, the better doth growe,
    and handsome ye make it at harvest to mowe.

Otes, barlie and pease, harrow after you sowe,
    for rie harrow first, as alreadie ye knowe.
Leave wheat little clod, for to cover the head,
    that after a frost, it may out and go spread.

If clod in thy wheat wil not breake with the frost,
    if now ye doo rowle it, it quiteth the cost.
But see when ye rowle it, the weather be drie,
    or else it were better unrowled to lie.

Gardening.   In March and in April, from morning to night,
    in sowing and setting, good huswives delight:
To have in a garden, or other like plot,
    to turn up their house, and to furnish their pot.

The nature of flowers dame Physick doth shew,
    she teacheth them all to be knowne to a few.
To set or to sowe, or else sowne to remove,
    how that should be practised, learne if ye love.

To know good land.   Land falling or lieng full South or southwest,
    for profit by tillage is lightly the best.
So garden with orchard and hopyard I finde,
    that want the like benefit, growe out of kinde.

If field to beare corne a good tillage doth crave,
    what thinke ye of garden, what garden would have?
In field without cost be assured of weedes,
    in garden be suer thou loosest thy seedes.

At spring (for the sommer) sowe garden ye shall,
    at harvest (for winter) or sowe not at all.
Oft digging, remooving, and weeding (ye see),
    makes herbe the more holesome and greater to bee.

Time faire, to sowe or to gather be bold,
  but set or remoove when the weather is cold.
Cut all thing or gather, the Moone in the wane,
  but sowe in encreasing, or give it his bane.

Now set doo aske watering with pot or with dish,
  new sowne doo not so, if ye doo as I wish.
Through cunning with dible, rake, mattock, and spade,
  by line and by leavell, trim garden is made.

Who soweth too lateward, hath seldome good seed,
  who soweth too soone, little better shall speed.
Apt time and the season so divers to hit,
  let aier and laier helpe practise and wit.

Now leekes are in season, for pottage full good,
  and spareth the milchcow and purgeth the blood.
These having, with peason for pottage in Lent,
  thou sparest both otemell and bread to be spent.

Though never so much a good huswife doth care,
  that such as doe labour have husbandlie fare.
Yet feed them and cram them til purse doe lack chinke,
  no spoone meat, no bellifull, labourers thinke.

Kill crowe, pie and cadow, rooke, buzard and raven,   *Destroie*
  or else go desire them to seeke a new haven.   *pie, rooks,*
In scaling the yoongest, to pluck off his beck,   *and ravens*
  beware how ye climber, for breaking your neck.   *nest, etc.*

Thus endeth Marches husbandrie.

APRILS ABSTRACT

Chapter 37

SOME champions laie
to fallow in Maie.

When tilth plows breake,
poore cattle cries creake.

One daie er ye plow,
spred compas ynow.

Some fodder buieth,
in fen where it lieth.

Thou champion wight,
have cow meat for night.

Set hop his pole,
make deepe the hole.

First, bark go and sell,
er timber ye fell.

Fence copie in,
er heawers begin.

The straightest ye knowe,
for staddles let growe.

Crab tree preserve,
for plough to serve.

Get timber out,
er yeere go about.

Som cuntries lack plowmeat,
and some doe want cowmeat.

Small commons and bare,
yeelds cattell ill fare.

Som common with geese,
and sheepe without fleese.
Som tits thither bring,
and hogs without ring.

Some champions agree
as waspe doth with bee.

Get swineherd for hog,
but kill not with dog.
Wher swineherd doth lack,
corne goeth to wrack.

All goes to the Devill,
where shepherd is evill.

Come home from land,
with stone in hand.

Man cow provides,
wife dairie guides.

Slut Cisley untaught
hath whitemeat naught.

Some bringeth in gaines,   Such Mistris, such Nan,
some losse beside paines.   such Maister, such Man.

Run Cisse, fault known,   *Thus endeth Aprils abstract, agree-*
with more than thine own.   *ing with Aprils husbandrie.*

APRILS HUSBANDRIE

Chapter 38

Sweete April showers,   Forgotten month past,
Doo spring Maie flowers.   Doe now at the last.

IN Cambridge shire forward to Lincolne shire way,
    the champion maketh his fallow in May.
Then thinking so dooing one tillage woorth twaine,
    by forcing of weede, by that meanes to refraine.

If April be dripping, then doo I not hate,
    (for him that hath little) his fallowing late,
Else otherwise fallowing timelie is best,
    for saving of cattel, of plough and the rest.

Be suer of plough to be readie at hand,
    er compas ye spred that on hillocks did stand:
Least drieng so lieng, doo make it decaie,
    er ever much water doo wash it awaie.

Looke now to provide ye of meadow for hay,
    if fennes be undrowned, there cheapest ye may.
In fen for the bullock, for horse not so well,
    count best the best cheape, wheresoever ye dwell.

Provide ye of cowmeate, for cattel at night,
　　and chiefly where commons lie far out of sight:
Where cattel lie tied without any meat,
　　that profit by dairie can never be great.

Put poles to　Get into thy hopyard with plentie of poles,
your hophils.　　　amongst those same hillocks devide them by doles.
　　　　　Three poles to a hillock (I pas not how long)
　　　　　　　shall yeeld thee more profit, set deeplie and strong.

Felling of　Sell barke to the tanner er timber yee fell,
timber.　　　cut lowe by the ground or else doo ye not well.
　　　　In breaking save crooked, for mill and for ships,
　　　　　and ever in hewing save carpenters chips.

　　　　First see it well fenced er hewers begin,
　　　　　then see it well stadled, without and within ;
　　　　Thus being preserved and husbandlie donne,
　　　　　shall sooner raise profit, to thee or thy sonne.

Stadling of　Leave growing for stadles the likest and best,
woods.　　　though seller and buier dispatched the rest.
　　　　In bushes, in hedgerowe, in grove, and in wood,
　　　　　this lesson observed is needfull and good.

　　　　Save elme, ash and crabtree, for cart and for plough,
　　　　　save step for a stile, of the crotch of the bough.
　　　　Save hazel for forks, save sallow for rake,
　　　　　save hulver and thorne, thereof flaile for to make.

Discharge　Make riddance of carriage, er yeere go about,
thy woods.　　　for spoiling of plant that is newlie come out.
　　　　To carter (with oxen) this message I bring,
　　　　　leave oxen abrode for anoieng the spring.

　　　　Allowance of fodder some countries doo yeeld,
　　　　　as good for the cattel as haie in the feeld.
　　　　Some mowe up their hedlonds and plots among corne,
　　　　　and driven to leave nothing, unmowne, or unshorne.

Some commons are barren, the nature is such,
  and some over laieth the common too much.
The pestered commons small profit doth geeve,
  and profit as little some reape I beleeve.

Some pester the commons, with jades and with geese,
  with hog without ring and with sheepe without fleese.
Some lose a daie labour with seeking their owne,
  some meet with a bootie they would not have knowne.

Great troubles and losses the champion sees,
  and ever in brauling, as wasps among bees:
As charitie that waie appeereth but small,
  so lesse be their winnings, or nothing at all.

Where champion wanteth a swineherd for hog,
  there many complaineth of naughtie mans dog.
Where ech his owne keeper appoints without care,
  there corne is destroied er men be aware.

The land is well harted with helpe of the fold,
  for one or two crops, if so long it will hold.
If shepherd would keepe them from stroieng of corne,
  the walke of his sheepe might the better be borne.

Where stones be too manie, annoieng thy land,
  make servant come home with a stone in his hand.
By daily so dooing, have plentie yee shall,
  both handsome for paving and good for a wall.

From April beginning, till Andrew be past,   *Dairie*
  so long with good huswife, hir dairie doth last.   *matters.*
Good milchcow and pasture, good husbands provide,
  the resdue good huswives knowes best how to guide.

Ill huswife unskilful to make hir owne chees,   *Ill hus-*
  through trusting of others hath this for hir fees.   *wiferie.*
Her milke pan and creame pot, so slabbered and sost,
  that butter is wanting and cheese is halfe lost.

Where some of a cow doo raise yeerelie a pound,
　with such seelie huswives no penie is found.
Then dairie maid (Cisley) hir fault being knowne,
　away apace trudgeth, with more than hir owne.

Ill huswives saiengs.

Then neighbour, for Gods sake, if any you see,
　good servant for dairie house, waine her to mee.
Such maister such man, and such mistris such maid,
　such husband and huswife, such houses araid.

A LESSON
for DAIRIE MAID CISLEY
OF TEN TOPPINGS GESTS

AS wife that will
　good husband plese,
Must shun with skill
　such gests as these.

So Cisse that serves
　must marke this note,
What fault deserves
　a brushed cote.

Ten toppings gests unsent for.

Gehezie, Lots wife, and Argusses eies,
Tom piper, poore Cobler, and Lazarus thies,
Rough Esau, with Mawdlin, and Gentils that scrall,
With Bishop that burneth, thus knowe ye them all.

> *These toppingly gests be in number but ten,*
> *As welcome in dairie as Beares among men.*
> *Which being descried, take heede of you shall,*
> *For danger of after claps, after that fall.*

Gehezie his sicknes was whitish and drie, / such cheeses, good Cisley, ye floted too nie.	*White and drie.*
Leave Lot with her piller (good Cisley) alone, / much saltnes in whitemeat is ill for the stone.	*Too salt.*
If cheeses in dairie have Argusses eies, / tell Cisley the fault in hir huswiferie lies.	*Full of eies.*
Tom Piper hath hoven and puffed up cheekes, / if cheese be so hoven, make Cisse to seeke creekes.	*Hoven.*
Poore Cobler he tuggeth his leatherlie trash, / if cheese abide tugging, tug Cisley a crash.	*Tough.*
If Lazer so lothsome in cheese be espied, / let baies amend Cisley, or shift hir aside.	*Full of spots.*
Rough Esau was hearie from top to the fut, / if cheese so appeareth, call Cisley a slut.	*Full of heares.*
As Mawdlin wept, so would Cisley be drest, / for whey in hir cheese, not halfe inough prest.	*Full of whey.*
If gentils be scrauling, call magget the py, / if cheeses have gentils, at Cisse by and by.	*Full of gentils.*
Blesse Cisley (good mistris) that Bishop doth ban / for burning the milke of hir cheese to the pan.	*Burnt to the pan.*

If thou (so oft beaten)   I will no more threaten,
  amendest by this.   I promise thee Cis.

Thus dairie maid Cisley, rehearsed ye see,
  what faults with ill huswife, in dairie house bee.
Of market abhorred, to houshold a griefe,
  to maister and mistris, as ill as a thiefe.

Thus endeth Aprils husbandrie.

MAIES ABSTRACT

Chapter 39

PUT lambe from eawe,
to milke a feawe.

Be not too bold,
to milke and to fold.

Five eawes alow,
to everie cow.

Sheepe wrigling taile
hath mads without faile.

Beat hard in the reede
where house hath neede.

Leave cropping from May
to Mihelmas day.
Let Ivie be killed,
else tree will be spilled.

Now threshers warne
to rid the barne.

Be suer of hay
till thend of May.

Let sheepe fill flanke,
where corne is too ranke.
In woodland lever,
in champion never.

To weeding away,
as soone as yee may.

For corne here reede,
what naughtie weede.

Who weeding slacketh,
good husbandrie lacketh.

Sowe buck or branke,
that smels so ranke.

Thy branke go and sowe,
where barlie did growe.
The next crop wheat
is husbandrie neat.

Sowe pescods some,
for harvest to come.

Sowe hemp and flacks.
that spinning lacks.

Teach hop to clime,
for now it is time.

Through fowles & weedes
poore hop ill speedes.
Cut off or crop
superfluous hop:
The titters or tine
makes hop to pine.

Some raketh their wheat,
with rake that is great.
So titters and tine
be gotten out fine.

Now sets doe crave
some weeding to have.

Now draine as ye like
both fen and dike.

Watch bees in May,
for swarming away.
Both now and in June,
marke maister bees tune.

Twifallow thy land,
least plough else stand.

No longer tarrie,
out compas to carrie.

Where neede doth pray it,
there see ye lay it.

Set Jack and Jone
to gather up stone.

To grasse with thy calves,
take nothing to halves.

Be suer thy neat
have water and meat.

By tainting of ground,
destruction is found.

Now carrege get
home fewell to fet.
Tell fagot and billet
for filching gillet.

In sommer for firing
let citie be buying.
Marke colliers packing
least coles be lacking.
(See opened sack)
for two in a pack.

Let nodding patch
go sleepe a snatch.

Wife as you will,
now plie your still.

Fine bazell sowe,
in a pot to growe.
Fine seedes sowe now,
before ye sawe how.

Keepe ox from cow,
for causes ynow.

*Thus endeth Maies abstract,
agreeing with Maies husbandrie.*

TWO OTHER SHORT
REMEMBRANCES

From bull cow fast
till Crowchmas be past.
From heifer bul hid thee
till Lammas doth bid thee.

103

MAIES HUSBANDRIE

Chapter 40

Cold Maie and windie,   *Forgotten month past,*
Barne filleth up finelie.   *Doe now at the last.*

Essex and   AT Philip and Jacob, away with the lams
Suffolke.   that thinkest to have any milke of their dams.
                At Lammas leave milking, for feare of a thing:
                  least (*requiem æternam*) in winter they sing.

Milking of   To milke and to fold them is much to require,
eawes.   except yee have pasture to fil their desire.
                Yet manie by milking (such heede they doo take),
                  not hurting their bodies much profit doo make.

                Five eawes to a cow, make a proofe by a score,
                  shall double thy dairie, else trust me no more.
                Yet may a good huswife that knoweth the skill,
                  have mixt and unmixt at hir pleasure and will.

                If sheepe or thy lambe fall a wrigling with taile,
                  go by and by search it, whiles helpe may prevaile:
                That barberlie handled I dare thee assure,
                  cast dust in his arse, thou hast finisht thy cure.

                Where houses be reeded (as houses have neede),
                  now pare off the mosse, and go beat in the reed.
                The juster ye drive it, the smoother and plaine,
                  more handsome ye make it to shut off the raine.

Leave off   From Maie til October leave cropping, for why?
cropping.   in wood sere, whatsoever thou croppest wil dy.
Destroie   Where Ivie imbraceth the tree verie sore,
Ivie.   kill Ivie, or else tree wil addle no more.

Keepe threshing for thresher, til Maie be come in,
   to have to be suer fresh chaffe in the bin.
And somewhat to scamble, for hog and for hen,
   and worke when it raineth for loitering men.

Be sure of haie and of provender some,   *Count store*
   for labouring cattel til pasture be come.   *no sore.*
And if ye doo mind to have nothing to sterve,
   have one thing or other, for all thing to serve.

Ground compassed wel and a following yeare,
   (if wheat or thy barlie too ranke doo appeare)
Now eat it with sheepe or else mowe it ye may,
   for ledging, and so, to the birds for a pray.

In Maie get a weede hooke, a crotch and a glove,   *Weeding.*
   and weed out such weedes as the corne doth not love:
For weeding of winter corne now it is best,
   but June is the better for weeding the rest.

The May weed doth burn and the thistle doth freat,   *Il weeds.*
   the fitchis pul downward, both rie and the wheat.
The brake and the cockle be noisome too much,
   yet like unto boddle no weede there is such.

Slack never thy weeding, for dearth nor for cheape,
   the corne shall reward it er ever ye reape.
And specially where ye doo trust for to seede,
   let that be well used, the better to speede.

In Maie is good sowing, thy buck or thy branke,   *Sowing of*
   that black is as pepper, and smelleth so ranke.   *branke.*
It is to thy land, as a comfort or muck,
   and al thing it maketh as fat as a buck.

Sowe buck after barlie, or after thy wheat,
   a peck to a roode (if the measure be great);
Three earthes see ye give it, and sowe it above,
   and harrow it finelie if buck ye doo love.

       Who pescods would gather, to have with the last,
         to serve for his houshold till harvest be past,
       Must sowe them in Maie, in a corner ye shal,
         where through so late growing no hindrance may fal.

Sowing of flax and hempe.

       Good flax and good hemp for to have of hir owne,
         in Maie a good huswife will see it be sowne.
       And afterward trim it, to serve at a neede,
         the fimble to spin and the karl for hir seede.

       Get into the hopyard, for now it is time,
         to teach Robin hop on his pole how to clime:
       To follow the Sunne, as his propertie is,
         and weede him and trim him, if aught go amis.

Il neighbours to the hop.

       Grasse, thistle and mustard seede, hemlock and bur,
         tine, mallow and nettle, that keepe such a stur.
       With peacock and turkie, that nibbles off top,
         are verie ill neighbors to seelie poore hop.

       From wheat go and rake out the titters or tine,
         if eare be not foorth, it will rise againe fine.
       Use now in thy rie, little raking or none,
         breake tine from his roote, and so let it alone.

Weeding of quickset.

       Bankes newly quicksetted, some weeding doo crave,
         the kindlier nourishment thereby to have.
       Then after a shower to weeding a snatch,
         more easilie weede with the roote to dispatch.

Now draine ditches.

       The fen and the quamire, so marrish be kind,
         and are to be drained, now wine to thy mind:
       Which yeerelie undrained and suffered uncut,
         annoieth the meadowes that thereon doo but.

Swarming of bees.

       Take heede to thy bees, that are readie to swarme,
         the losse thereof now is a crownes worth of harme:
       Let skilfull be readie and diligence seene,
         least being too careles, thou losest thy beene.

In Maie at the furthest, twifallow thy land,   *Twifallow-*
   much drout may else after cause plough for to stand: *ing.*
This tilth being done, ye have passed the wurst,
   then after who ploweth, plow thou with the furst.

Twifallow once ended, get tumbrell and man,   *Carie out*
   and compas that fallow as soone as ye can.   *compas.*
Let skilfull bestow it, where neede is upon,
   more profit the sooner to follow thereon.

Hide hedlonds with muck, if ye will to the knees,
   so dripped and shadowd with bushes and trees:
Bare plots full of galles, if ye plow overthwart,
   and compas it then, is a husbandlie part.

Let children be hired, to lay to their bones,
   from fallow as needeth to gather up stones.
What wisedome for profit adviseth unto,
   that husband and huswife must willingly do.

To gras with thy calves in some medow plot nere,   *Forth to*
   where neither their mothers may see them nor here.   *grasse with*
Where water is plentie and barth to sit warme,   *thy calves.*
   and looke well unto them, for taking of harme.

Pinch never thy wennels of water or meat,   *Let not*
   if ever ye hope for to have them good neat: *cattel want*
In Sommer time dailie, in Winter in frost,   *water.*
   if cattel lack drinke, they be utterly lost.

For coveting much overlay not thy ground,   *Overlay*
   and then shall thy cattel be lustie and sound.   *not thy*
But pinch them of pasture, while Sommer doth last,   *pastures.*
   and lift at their tailes er an Winter be past.

Get home with thy fewell, made readie to fet,   *Get home*
   the sooner the easier carrege to get: *thy fewel.*
Or otherwise linger the carrege thereon,
   till (where as ye left it) a quarter be gon.

Husbandrie   His firing in Sommer, let Citizen buie,
for Citizens.   least buieng in Winter make purse for to crie.
                  For carman and collier harps both on a string,
                     in Winter they cast to be with thee to bring.

Sleeping   From Maie to mid August, an hower or two,
time.   let patch sleepe a snatch, how soever ye do,
                  Though sleeping one hower refresheth his song,
                     yet trust not hob growthed for sleeping too long.

Stilling   The knowledge of stilling is one pretie feat,
of herbes.   The waters be holesome, the charges not great.
                  What timelie thou gettest, while Sommer doth last,
                     thinke Winter will helpe thee, to spend it as fast.

                  Fine bazell desireth it may be hir lot,
                     to growe as the gilloflower, trim in a pot,
                  That ladies and gentils, for whom she doth serve,
                     may helpe hir as needeth, poore life to preserve.

                  Keepe oxe fro thy cow that to profit would go,
                     least cow be deceived by oxe dooing so :
                  And thou recompenced for suffering the same,
                     with want of a calfe and a cow to wax lame.

                      Thus endeth Maies husbandrie.

JUNES ABSTRACT

Chapter 41

WASH sheep for to share,
that sheepe may go bare.

Though fleese ye take,
no patches make.

Share lambes no whit,
or share not yit.

If meadow be growne,
let meadow be mowne.

Plough early ye may,
and then carrie hay.

Tis good to be knowne,
to have all of thine owne.
Who goeth a borrowing,
goeth a sorrowing.

See cart in plight,
and all things right.

Make drie over hed,
both hovell and shed.

Of hovell make stack,
for pease on his back.

In champion some,
wants elbow rome.

Let wheat and rie,
in house lie drie.

Buie turfe and sedge,
or else breake hedge.

Good store howse needfull
well ordred speedfull.

Thy barnes repaire,
make flower faire.

Such shrubs as noie,
in sommer destroie.

Swinge brembles & brakes,
get forkes and rakes.

Spare hedlonds some,
till harvest come.

Cast ditch and pond,
to lay upon lond.

A LESSON OF HOPYARD

Where hops will growe,
here learne to knowe.
Hops many will coome,
in a roode of roome.

Hops hate the land,   Hop plot once found,
with gravell and sand.   now dig the ground.

The rotten mold   Hops favoreth malt,
for hop is worth gold.   hops thrift doth exalt:
                               Of hops more reede,
The sunne southwest   as time shall neede.
for hopyard is best.

Thus endeth Junes abstract, agreeing with Junes husbandrie.

JUNES HUSBANDRIE

Chapter 42

Calme weather in June   *Forgotten month past,*
Corne sets in tune.   *Doe now at the last.*

Sheepe sharing.   WASH sheepe (for the better) where water doth run,
        and let him go cleanly and drie in the sun.
   Then share him and spare not, at two daies an end,
        The sooner the better his corps will amend.

Beware of evill sheepe shearers.   Reward not thy sheepe (when ye take off his cote)
        with twitchis and patches, as brode as a grote.
   Let not such ungentlenesse happen to thine,
        least flie with hir gentils doo make it to pine.

Sheare lambes in Julie.   Let lambes go unclipped, till June be halfe worne,
        the better the fleeses will growe to be shorne.
   The Pie will discharge thee for pulling the rest:
        the lighter the sheepe is, then feedeth it best.

If meadow be forward, be mowing of some;   *Mowing*
  but mowe as the makers may well overcome: *time.*
Take heede to the weather, the wind and the skie,
  if danger approcheth, then cock apace crie.

Plough earlie till ten a clock, then to thy hay,
  in plowing and carting, so profit ye may.
By little and little, thus dooing ye win:
  that plough shall not hinder when harvest comes in.

Provide of thine owne to have all things at hand,
  least worke and the workman unoccupide stand.
Love seldome to borowe that thinkest to save,
  for he that once lendeth twise looketh to have.

Let cart be well searched without and within,   *Trim wel*
  well clouted and greased, er hay time begin.   *thy carts.*
Thy hay being carried, though carter had sworne,
  carts bottome well boorded is saving of corne.

Good husbands that laie to save all things upright,
  for tumbrels and cart, have a shed readie dight.
Where under the hog may in winter lie warme:
  to stand so enclosed, as wind doo no harme.

So likewise a hovell will serve for a roome,   *A hovell is set*
  to stack on the peason, when harvest shall coome.   *upon crotches &*
And serve thee in winter, more over than that,   *covered with*
  to shut up thy porklings thou mindest to fat.   *poles & strawe.*

Some barnroome have little, and yardroome as much,
  yet corne in the field appertaineth to such:
Then hovels and rikes they are forced to make,
  abrode or at home for necessities sake.

Make sure of breadcorne (of all other graine),
  lie drie and well looked to, for mouse and for raine.
Though fitchis and pease, and such other as they,
  (for pestring too much) on a hovell ye ley.

With whinnes or with furzes thy hovell renew,
   for turfe or for sedge, for to bake and to brew:
For charcole and sea cole, as also for thacke,
   for tallwood and billet, as yeerlie ye lacke.

The husbandlie storhouse.

What husbandlie husbands, except they be fooles,
   but handsome have storehouse, for trinkets and tooles:
And all in good order, fast locked to ly,
   what ever is needfull, to find by and by.

Thy houses and barnes would be looked upon,
   and all things amended er harvest come on.
Things thus set in order, in quiet and rest,
   shall further thy harvest and pleasure thee best.

The bushes and thorne with the shrubs that do noy,
   in woodsere or sommer cut downe to destroy:
But where as decay to the tree ye will none,
   for danger in woodsere, let hacking alone.

Mowe downe brakes and meadow.

At Midsommer, downe with the brembles and brakes,
   and after, abrode with thy forks and thy rakes:
Set mowers a mowing, where meadow is growne,
   the longer now standing the worse to be mowne.

Mowe hedlonds at harvest or after in the several fields.

Now downe with the grasse upon hedlonds about,
   that groweth in shadow, so ranke and so stout.
But grasse upon hedlond of barlie and pease,
   when harvest is ended, go mowe if ye please.

Such muddie deepe ditches, and pits in the feeld,
   that all a drie sommer no water will yeeld,
By fieing and casting that mud upon heapes,
   commodities many the husbandman reapes.

A LESSON
WHERE AND WHEN
to plant good Hopyard.

Whome fancie persuadeth, among other crops,
   to have for his spending, sufficient of hops,
Must willinglie follow, of choises to chuse,
   such lessons approoved, as skilfull doo use.

Ground gravellie, sandie, and mixed with clay,   *Naught*
   is naughtie for hops any maner of way;   *for hops.*
Or if it be mingled with rubbish and stone,
   for drines and barrennnes, let it alone.

Choose soile for the hop of the rottenest mould,   *Good for*
   well doonged and wrought, as a garden plot should: *hops.*
Not far from the water (but not overflowne)
   this lesson well noted is meete to be knowne.

The Sunne in the south, or else southly and west,
   is joy to the hop, as a welcomed gest;
But wind in the north, or else northly east,
   to hop is as ill as a fraie in a feast.

Meete plot for a hopyard once found as is told,   *Now dig*
   make thereof account, as of jewell of gold.   *thy new hop*
Now dig it and leave it, the Sunne for to burne,   *ground.*
   and afterward fence it, to serve for that turne.

The hop for his profit I thus doo exalt,   *The praise*
   it strengtheneth drinke, and it favoreth malt.   *of hops.*
And being well brewed, long kept it will last,
   and drawing abide, if ye drawe not too fast.

JULIES ABSTRACT

Chapter 43

GO, sirs, and away,
to ted and make hay.
If stormes drawes nie,
then cock apace crie.

Let hay still bide,
till well it be dride.
(Hay made) away carrie,
no longer then tarrie.

Who best way titheth,
he best way thriveth.

Two good hay makers
woorth twentie crakers.

Let dallops about
be mowne and had out.
See hay doo looke greene,
see feeld ye rake cleene.

Thry fallow I pray thee,
least thistles bewray thee.

Cut off, good wife,
ripe beane with a knife.

Ripe hempe out cull,
from karle to pull.
Let seede hempe growe,
till more ye knowe.

Drie flax get in,
for spinners to spin.
Now mowe or pluck
thy branke or buck.

Some wormewood save,
for March to have.

Mark Physick true,
of wormewood and rue.
Get grist to the mill,
for wanting at will.

*Thus endeth Julies
abstract, agreeing with
Julies husbandrie.*

JULIES HUSBANDRIE

Chapter 44

No tempest, good Julie, *Forgotten month past,*
Least corne lookes rulie. *Doe now at the last.*

GO muster thy servants, be captaine thy selfe,   *Hay*
  providing them weapon and other like pelfe.   *harvest.*
Get bottles and walletts, keepe field in the heat,
  the feare is as much, as the danger is great.

With tossing and raking and setting on cox,
  grasse latelie in swathes is hay for an ox:
That done, go and cart it and have it away,
  the battel is fought, ye have gotten the day.

Pay justly thy tithes whatsoever thou bee,   *Pay thy*
  that God may in blessing send foison to thee.   *tithes.*
Though Vicar be bad, or the Parson as evill,
  go not for thy tithing thy selfe to the Devill.

Let hay be well made, or avise else avouse,
  for molding in goef, or of firing the house.
Lay coursest aside for the ox and the cow,
  the finest for sheepe and thy gelding alow.

Then downe with the hedlonds, that groweth about,
  leave never a dallop unmowne and had out.
Though grasse be but thin, about barlie and pease,
  yet picked up cleane ye shall find therein ease.

Thry fallow betime, for destroieng of weede,   *Thry fal-*
  least thistle and duck fall a blooming and seede,   *lowing.*
Such season may chance, it shall stand thee upon,
  to till it againe, er an Sommer be gon.

Gathering of garden beanes.

Not rent off, but cut off, ripe beane with a knife,
   for hindering stalke of hir vegetive life.
So gather the lowest, and leaving the top,
   shall teach thee a trick, for to double thy crop.

Gather yellow hempe.

Wife, pluck fro thy seed hemp the fiemble hemp clene,
   this looketh more yellow, the other more grene:
Use ton for thy spinning, leave Mihel the tother,
   for shoo thred and halter, for rope and such other.

Now pluck up thy flax, for the maidens to spin,
   first see it dried, and timelie got in.
And mowe up thy branke, and away with it drie,
   and howse it up close, out of danger to lie.

Wormewood get against fleas and infection.

While wormwood hath seed, get a handful or twaine,
   to save against March to make flea to refraine:
Where chamber is sweeped, and wormwood is strowne,
   no flea for his life dare abide to be knowne.

What saver is better (if physick be true),
   for places infected, than wormwood and rue.
It is as a comfort for hart and the braine,
   and therefore to have it, it is not in vaine.

Be sure of bread and drinke for harvest.

Get grist to the mill, to have plentie in store,
   least miller lack water, as many doo more.
The meale the more yeeldeth, if servant be true,
   and miller that tolleth, take none but his due.

Thus endeth Julies husbandrie.

AUGUSTS ABSTRACT

Chapter 45

THRY fallowing won,
get compassing don.

In June and in Awe
swinge brakes (for a lawe).

Pare saffron plot,
forget it not.
His dwelling made trim,
looke shortly for him:
When harvest is gon,
then saffron comes on.

A little of ground
brings saffron a pound.
The pleasure is fine,
the profit is thine.
Keepe colour in drieng,
well used woorth buieng.

Maids, mustard seed reape,
and laie on a heape.

Good neighbors in deede,
change seede for seede.

Now strike up drum,
cum harvest man cum.
Take paine for a gaine,
one knave mars twaine.

Reape corne by the day,
least corne doo decay.
By great is the cheaper,
if trustie were reaper.

Blowe horne for sleapers,
and cheere up thy reapers.

Well dooings who loveth,
thes harvest points proveth.

Paie Gods part furst,
and not of the wurst.

Now Parson (I say),
tith carrie away.

Keepe cart gap weele,
scare hog from wheele.

Mowe hawme to burne,
to serve thy turne:
To bake thy bread,
to burne under lead.

Mowne hawme being dry,
no longer let ly.
Get home thy hawme,
whilst weather is cawme.

Mowne barlie lesse cost,
ill mowne much lost.

Reape barlie with sickle,
that lies in ill pickle.
Let greenest stand,
for making of band.
Bands made without dew,
will hold but a few.

Laie band to find her,
two rakes to a binder.

Rake after sieth,
and pay thy tieth,
Corne carried all,
then rake it ye shall.

Let shock take sweate,
least gofe take heate.
Yet it is best reason,
to take it in season.

More often ye turne,
more pease ye out spurne.
Yet winnow them in,
er carrege begin.

Thy carting plie,
while weather is drie.

Bid goving (clim)
gove just and trim.
Laie wheat for seede,
to come by at neede.

Seede barelie cast,
to thresh out last.

Lay pease upon stacke,
if hovell ye lack.
And cover it straight,
from doves that waight.

Let gleaners gleane,
(the poore I meane).
Which ever ye sowe,
that first eate lowe.
The other forbare,
for rowen to spare.

Come home lord singing,
com home corne bringing.
Tis merie in hall,
when beards wag all.

Once had thy desire,
pay workman his hire.
Let none be beguilde,
man, woman, nor childe.

Thanke God ye shall,
and adue for all.

WORKS AFTER HARVEST

Get tumbrell in hand,
for barlie land.

The better the muck,
the better good luck.

Still carrege is good,
for timber and wood.

No longer delaies,
to mend the high waies.

Some love as a jewell,
well placing of fewell.

In piling of logs,
make hovell for hogs.

Wife, plow doth crie,
to picking of rie.

Such seede as ye sowe,
such reape or else mowe.

Take shipping or ride.
Lent stuffe to provide.

Let haberden lie,
in peasestraw drie.

When out ye ride,
leave a good guide.

Some profit spie out,
by riding about.
Marke now, thorow yeere,
what cheape, what deere.

Some skill doth well
to buie and to sell.
Of theefe who bieth,
in danger lieth.

Commoditie knowne,
abrode is blowne.

At first hand bie,
at third let lie.

Have monie prest,
to buie at the best.

Some cattle home bring,
for Mihelmas spring.
By hauke and hound,
small profit is found.

Dispatch, looke home,
to loitring mome.
Provide or repent,
milch cow for Lent.

Now crone your sheepe,
fat those ye keepe.
Leave milking old cow,
fat aged up now.

Sell butter and cheese,
good Faires few leese.
At Faires go bie,
home wants to supplie.

If hops looke browne,
go gather them downe.
But not in the deaw,
for piddling with feaw.

Of hops this knack,
a meanie doo lack.
Once had thy will,
go cover his hill.

Take hop to thy dole,
but breake not his pole.

Learne here (thou stranger)
to frame hop manger.

Hop poles preserve,
againe to serve.
Hop poles by and by,
long safe up to dry.
Least poles wax scant,
new poles go plant.

The hop kell dride,
will best abide.

Hops dried in loft,
aske tendance oft.
And shed their seedes,
much more than needes.

Hops dride small cost,
ill kept halfe lost.
Hops quickly be spilt,
take heede if thou wilt.

Some come, some go,
This life is so.

*Thus endeth Augusts
abstract, agreeing with
Augusts husbandrie.*

AUGUSTS HUSBANDRIE

Chapter 46

Drie August and warme,   *Forgotten month past,*
Doth harvest no harme.   *Doe now at the last.*

THRY fallow once ended, go strike by and by,   *Thry fallow-*
   both wheat land and barlie, and so let it ly.   *ing.*
And as ye have leisure, go compas the same,
   when up ye doo lay it, more fruitfull to frame.

Get downe with thy brakes, er an showers doo come,   *Mowing of*
   that cattle the better may pasture have some.   *brakes.*
In June and in August, as well doth appeere,
   is best to mowe brakes, of all times in the yeere.

Pare saffron betweene the two S. Maries daies,   *Paring of*
   or set or go shift it, that knowest the waies.   *saffron.*
What yeere shall I doo it (more profit to yeeld?)
   the fourth in garden, the third in the feeld.

In having but fortie foote workmanly dight,   *Huswiferie.*
   take saffron ynough for a Lord and a knight.
All winter time alter as practise doth teach,
   what plot have ye better, for linnen to bleach.

Maides, mustard seede gather, for being too ripe,
   and weather it well, er ye give it a stripe:
Then dresse it and laie it in soller up sweete,
   least foistines make it for table unmeete.

Good huswifes in sommer will save their owne seedes,
   against the next yeere, as occasion needes.
One seede for another, to make an exchange,
   with fellowlie neighbourhood seemeth not strange.

Corne
harvest.

Make sure of reapers, get harvest in hand,
   the corne that is ripe, doo but shed as it stand.
Be thankfull to God, for his benefits sent,
   and willing to save it with earnest intent.

Champion
by great,
the other
by day.

To let out thy harvest, by great or by day,
   let this by experience leade thee a way.
By great will deceive thee, with lingring it out,
   by day will dispatch, and put all out of dout.

Grant harvest lord more by a penie or twoo,
   to call on his fellowes the better to doo:
Give gloves to thy reapers, a larges to crie,
   and dailie to loiterers have a good eie.

Good
harvest
points.

Reape wel, scatter not, gather cleane that is shorne,
   binde fast, shock apace, have an eie to thy corne.
Lode safe, carrie home, follow time being faire,
   gove just in the barne, it is out of despaire.

Tithe dulie and trulie, with hartie good will,
   that God and his blessing may dwell with thee still:
Though Parson neglecteth his dutie for this,
   thanke thou thy Lord God, and give erie man his.

Parson
looke to
thy tithe.

Corne tithed (sir Parson) to gather go get,
   and cause it on shocks to be by and by set:
Not leaving it scattering abrode on the ground,
   nor long in the field, but away with it round.

Keepe hog
from cart
wheele.

To cart gap and barne, set a guide to looke weele,
   and hoy out (sir carter) the hog fro thy wheele:
Least greedie of feeding, in following cart,
   it noieth or perisheth, spight of thy hart.

In champion countrie a pleasure they take,
   to mowe up their hawme, for to brew and to bake.
And also it stands them in steade of their thack,
   which being well inned, they cannot well lack.

The hawme is the strawe of the wheat or the rie,
  which once being reaped, they mowe by and bie:
For feare of destroieng with cattle or raine,
  the sooner ye lode it, more profit ye gaine.

The mowing of barlie, if barlie doo stand, *Mowing of*
  is cheapest and best, for to rid out of hand: *barlie.*
Some mowe it and rake it, and sets it on cocks,
  some mowe it and binds it, and sets it on shocks.

Of barlie the longest and greenest ye find, *Binding of*
  leave standing by dallops, till time ye doo bind: *barlie.*
Then early in morning (while deaw is thereon),
  to making of bands till the deaw be all gon.

One spreadeth those bands, so in order to ly, *Spreading*
  as barlie (in swatches) may fill it thereby: *of barlie*
Which gathered up, with the rake and the hand, *bands.*
  the follower after them bindeth in band.

Where barlie is raked (if dealing be true), *Tithe of*
  the tenth of such raking to Parson is due: *rakings.*
Where scatring of barlie is seene to be much,
  there custome nor conscience tithing should gruch.

Corne being had downe (any way ye alow),
  should wither as needeth, for burning in mow:
Such skill apperteineth to harvest mans art,
  and taken in time is a husbandly part.

No turning of peason till carrege ye make, *Usage of*
  nor turne in no more, than ye mind for to take: *peason.*
Least beaten with showers so turned to drie,
  by turning and tossing they shed as they lie.

If weather be faire, and tidie thy graine, *Lingring*
  make speedily carrege, for feare of a raine: *lubbers.*
For tempest and showers deceiveth a menie,
  and lingering lubbers loose many a penie.

Best maner of goving corn in the barn.

In goving at harvest, learne skilfully how
   ech graine for to laie, by it selfe on a mow:
Seede barlie the purest, gove out of the way,
   all other nigh hand gove as just as ye may.

Pease stack.

Stack pease upon hovell abrode in the yard,
   to cover it quicklie, let owner regard:
Least Dove and the cadow, there finding a smack,
   with ill stormie weather doo perish thy stack.

Leave gleaning for the poore.

Corne carred, let such as be poore go and gleane,
   and after, thy cattle to mowth it up cleane.
Then spare it for rowen, till Mihel be past,
   to lengthen thy dairie no better thou hast.

In harvest time, harvest folke, servants and all,
   should make all togither good cheere in the hall:
And fill out the black boule of bleith to their song,
   and let them be merie all harvest time long.

Pay trulie harvest folke.

Once ended thy harvest, let none be begilde,
   please such as did helpe thee, man, woman, and childe.
Thus dooing, with alway such helpe as they can,
   thou winnest the praise of the labouring man.

Thanke God for all.

Now looke up to Godward, let tong never cease
   in thanking of him, for his mightie encrease:
Accept my good will, for a proofe go and trie:
   the better thou thrivest, the gladder am I.

WORKS AFTER HARVEST

Now carrie out compas, when harvest is donne,
   where barlie thou sowest, my champion sonne:
Or laie it on heape, in the field as ye may,
   till carriage be faire, to have it away.

Whose compas is rotten and carried in time,
   and spred as it should be, thrifts ladder may clime.
Whose compas is paltrie and carried too late,
   such husbandrie useth that many doo hate.

Er winter preventeth, while weather is good,   *Carriage of*
   for galling of pasture get home with thy wood.   *fewell.*
And carrie out gravell to fill up a hole:
   both timber and furzen, the turfe and the cole.

Howse charcole and sedge, chip and cole of the land,   *Well placing*
   pile tallwood and billet, stacke all that hath band.   *of fewell.*
Blocks, rootes, pole and bough, set upright to the thetch:
   the neerer more handsome in winter to fetch.

In stacking of baven, and piling of logs,   *Hovell for*
   make under thy baven a hovell for hogs,   *hogs.*
And warmelie enclose it, all saving the mouth,
   and that to stand open, and full to the south.

Once harvest dispatched, get wenches and boies,
   and into the barne, afore all other toies.
Choised seede to be picked and trimlie well fide,
   for seede may no longer from threshing abide.

Get seede aforehand, in a readines had,
   or better provide, if thine owne be too had.
Be carefull of seede, or else such as ye sowe,
   be sure at harvest, to reape or to mowe.

When harvest is ended, take shipping or ride,   *Provision*
   Ling, Saltfish and Herring, for Lent to provide.   *for Lent.*
To buie it at first, as it commeth to rode,
   shall paie for thy charges thou spendest abrode.

Choose skilfullie Saltfish, not burnt at the stone,
   buie such as be good, or else let it alone.
Get home that is bought, and goe stack it up drie,
   with peasestrawe betweene it, the safer to lie.

Compassing of barlie land.

Er ever ye jornie, cause servant with speede
   to compas thy barlie land where it is neede.
One aker well compassed, passeth some three,
   thy barne shall at harvest declare it to thee.

This lesson is learned by riding about,
   the prices of vittels, the yeere thorough out.
Both what to be selling and what to refraine,
   and what to be buieng, to bring in againe.

Though buieng and selling doth woonderfull well,
   to such as have skill how to buie and to sell:
Yet chopping and changing I cannot commend,
   with theefe and his marrow, for feare of ill end.

The rich in his bargaining needes not be tought,
   of buier and seller full far is he sought.
Yet herein consisteth a part of my text,
   who buieth at first hand, and who at the next.

Buieng at first hand.

At first hand he buieth that paieth all downe,
   at second, that hath not so much in the towne,
At third hand he buieth that buieth of trust,
   at his hand who buieth shall paie for his lust.

Readie monie bieth best cheape.

As oft as ye bargaine, for better or wurse,
   to buie it the cheaper, have chinkes in thy purse:
Touch kept is commended, yet credit to keepe,
   is paie and dispatch him, er ever ye sleepe.

Hauking.

Be mindfull abrode of Mihelmas spring,
   for thereon dependeth a husbandlie thing:
Though some have a pleasure, with hauke upon hand,
   good husbands get treasure, to purchase their land.

Winter milch cow.

Thy market dispatched, turne home againe round,
   least gaping for penie, thou loosest a pound:
Provide for thy wife, or else looke to be shent,
   good milch cow for winter, another for Lent.

In traveling homeward, buie fort ie good crones, / and fat up the bodies of those seelie bones. / Leave milking and drie up old mulley thy cow, / the crooked and aged, to fatting put now.	*Old ewes.*
At Bartilmewtide, or at Sturbridge faire, / buie that as is needfull, thy house to repaire: / Then sell to thy profit, both butter and cheese, / who buieth it sooner, the more he shall leese.	*Buieng or selling of butter and cheese.*
If hops doo looke brownish, then are ye too slowe, / if longer ye suffer those hops for to growe. / Now sooner ye gather, more profit is found, / if weather be faire and deaw of a ground.	*Hops gathering.*
Not breake off, but cut off, from hop the hop string, / leave growing a little againe for to spring. / Whose hill about pared, and therewith new clad, / shall nourish more sets against March to be had.	*Increasing of hops.*
Hop hillock discharged of everie let, / see then without breaking, ech pole ye out get. / Which being untangled above in the tops, / go carrie to such as are plucking of hops.	*The order of hops gathering.*
Take soutage or haier (that covers the kell), / set like to a manger and fastened well: / With poles upon crotchis as high as thy brest, / for saving and riddance is husbandrie best.	*Hop manger.*
Hops had, the hop poles that are likelie preserve, / (from breaking and rotting) againe for to serve: / And plant ye with alders or willowes a plot, / where yeerelie as needeth mo poles may be got.	*Save hop poles.*
Some skilfullie drieth their hops on a kell, / and some on a soller, oft turning them well. / Kell dried will abide, foule weather or faire, / where drieng and lieng in loft doo dispaire.	*Drieng of hops.*

*Keeping
of hops.*
    Some close them up drie in a hogshed or fat,
        yet canvas or soutage is better than that:
    By drieng and lieng they quickly be spilt:
        thus much have I shewed, doo now as thou wilt.

    Old fermer is forced long August to make,
        his goodes at more leisure away for to take.
    New fermer he thinketh ech houre a day,
        untill the old fermer be packing away.

    *Thus endeth and holdeth out Augusts husbandrie,
        till Mihelmas Eve.*
                              Tho. Tusser.

CORNE HARVEST
equally devided into
ten partes.

Chapter 47

    One part cast forth, for rent due out of hand,
    One other part, for seede to sowe thy land.
    Another part, leave Parson for his tieth.
    Another part for harvest, sickle and sieth.

For naperie,   One part for ploughwrite, cartwrite, knacker, smith,
sope and candle,   One part to uphold thy teemes that drawe therewith.
salt and sauce,   One part for servant and workmans wages lay.
tinker and   One part likewise for filbellie day by day.
cooper, brasse   One part thy wife for needfull things doth crave.
and pewter.   Thy selfe and childe, the last one part would have.

Who minds to cote,   Yet fermer may
  upon this note,   thanke God and say,
    may easily find ynough: for yeerlie such good hap:
What charge and paine,   Well fare the plough,
  to litle gaine,   that sends ynough
    doth follow toiling plough.   to stop so many a gap.

A BRIEFE CONCLUSION
where you may see, Ech word in the verse,
to begin with a T.

Chapter 48

THE thriftie that teacheth the thriving to thrive,   *Trive for*
Teach timelie to traverse the thing that thou trive.   *contrive.*
Transferring thy toiling, to timelines tought.
This teacheth thee temprance, to temper thy thought.

Take trustie (to trust to) that thinkest to thee,
That trustily thriftines trowleth to thee.
Then temper thy travell to tarie the tide,
This teacheth thee thriftines twentie times tride.

Take thankfull thy talent, thanke thankfully those
That thriftilie teacheth thy time to transpose.
Troth twise to thee teached, teach twentie times ten.
This trade thou that takest, take thrift to thee then.

Mans age divided here ye have,
By prentiships,
from birth to his grave.

Chapter 49

 7   *The first seven yeers bring up as a childe,*
14   *The next to learning, for waxing too wilde.*
21   *The next keepe under sir hobbard de hoy,*
28   *The next a man no longer a boy.*
35   *The next, let lustie laie wisely to wive,*
42   *The next, laie now or else never to thrive.*
49   *The next, make sure for terme of thy life,*
56   *The next, save somewhat for children and wife.*
63   *The next, be staied, give over thy lust,*
70   *The next, thinke hourely whither thou must.*
77   *The next, get chaire and crotches to stay,*
84   *The next, to heaven God send us the way.*

Who looseth their youth, shall rue it in age:
Who hateth the truth, in sorowe shall rage.

Another division of the nature
of mans age.

Chapter 50

The Ape, the Lion, the Foxe, the Asse,
Thus sets foorth man, as in a glasse.

APE	*Like Apes we be toieng, till twentie and one,*
LYON	*Then hastie as Lions till fortie be gone:*
FOXE	*Then wilie as Foxes, till threescore and three,*
ASSE	*Then after for Asses accounted we bee.*

Who plaies with his better, this lesson must knowe,
   what humblenes Foxe to the Lion doth owe.
Foxe, Ape with his toieng and rudenes of Asse,
   brings (out of good hower) displeasure to passe.

Comparing good husband with
unthrift his brother,
The better discerneth the tone
from the tother.

Chapter 51

Ill husbandrie braggeth,
  to go with the best:
Good husbandrie baggeth
  up gold in his chest.

Ill husbandry trudgeth,
  with unthrifts about:
Good husbandry snudgeth,
  for fear of a dout.

Ill husbandrie spendeth
  abrode like a mome:
Good husbandrie tendeth
  his charges at home.

Ill husbandrie selleth
  his corne on the ground:
Good husbandrie smelleth
  no gain that way found.

Ill husbandrie loseth,
  for lack of good fence:
Good husbandrie closeth,
  and gaineth the pence.

Ill husbandrie trusteth
  to him and to hur:
Good husbandrie lusteth
  himselfe for to stur.

Ill husbandrie eateth
  himselfe out a doore:
Good husbandrie meateth
  his friend and the poore.

Ill husbandrie daieth,
  or letteth it lie:
Good husbandrie paieth,
  the cheaper to bie.

Ill husbandrie lurketh,
  and stealeth a sleepe:
Good husbandrie worketh,
  his houshold to keepe.

Ill husbandrie liveth,
  by that and by this:
Good husbandrie giveth
  to erie man his.

Ill husbandrie taketh,
   and spendeth up all:
Good husbandrie maketh
   good shift with a small.

Ill husbandry praieth
   his wife to make shift:
Good husbandrie saieth
   take this of my gift.

Ill husbandry drowseth
   at fortune so auke:
Good husbandrie rowseth
   himselfe as a hauke.

Ill husbandrie lieth
   in prison for debt:
Good husbandrie spieth
   where profit to get.

Ill husbandrie waies
   has to fraud what he can:
Good husbandrie praies
   hath of everie man.

Ill husbandrie never
   hath welth to keep touch:
Good husbandrie ever
   hath penie in pouch.

    Good husband his boone,
      Or request hath a far.
    Ill husband assoone
      Hath a tode with an R.

A COMPARISON BETWEENE
Champion countrie and severall.

Chapter 52

                THE countrie enclosed I praise,
                   the tother delighteth not me,
                For nothing the wealth it doth raise,
                   to such as inferior be.
                How both of them partly I knowe,
                   here somewhat I mind for to showe.

Champion.   There swineherd that keepeth the hog,
                   there neatherd, with cur and his horne,
                There shepherd with whistle and dog,
                   be fence to the medowe and corne.
                There horse being tide on a balke,
                   is readie with theefe for to walke.

                Where all thing in common doth rest,
                   corne field with the pasture and meade,
                Though common ye doo for the best,
                   yet what doth it stand ye in steade?
                There common as commoners use,
                   for otherwise shalt thou not chuse.

                What laier much better then there,
                   or cheaper (thereon to doo well?)
                What drudgerie more any where
                   lesse good thereof where can ye tell?
                What gotten by Sommer is seene:
                   in Winter is eaten up cleene.

Example by Leicester shire,
    what soile can be better than that?
For any thing hart can desire,
    and yet doth it want ye see what.
Mast, covert, close pasture, and wood,
    and other things needfull as good.

All these doo enclosure bring,   *Enclosure.*
    experience teacheth no lesse,
I speake not to boast of the thing,
    but onely a troth to expresse.
Example (if doubt ye doo make):
    by Suffolke and Essex go take.

More plentie of mutton and biefe,   *Severall.*
    corne, butter, and cheese of the best,
More wealth any where (to be briefe),
    more people, more handsome and prest,
Where find ye? (go search any coast)
    than there where enclosure is most.

More worke for the labouring man,
    as well in the towne as the feeld:
Or thereof (devise if ye can)
    more profit what countries doo yeeld?
More seldome where see ye the poore,
    go begging from doore unto doore?

In Norfolke behold the dispaire   *Champion*
    of tillage too much to be borne: *countrie.*
By drovers from faire to faire,
    and others destroieng the corne.
By custome and covetous pates,
    by gaps, and by opening of gates.

What speake I of commoners by,
  with drawing all after a line:
So noieng the corne, as it ly,
  with cattle, with conies, and swine.
When thou hast bestowed thy cost,
  looke halfe of the same to be lost.

The flocks of the Lords of the soile
  do yeerly the winter corne wrong:
The same in a manner they spoile,
  with feeding so lowe and so long.
And therefore that champion feeld
  doth seldome good winter corne yeeld.

Champion noiances.

By Cambridge a towne I doo knowe,
  where many good husbands doo dwell;
Whose losses by losels doth showe,
  more here than is needfull to tell:
Determine at court what they shall,
  performed is nothing at all.

The champion robbeth by night,
  and prowleth and filcheth by day:
Himselfe and his beast out of sight,
  both spoileth and maketh away
Not onely thy grasse, but thy corne,
  both after, and er it be shorne.

Pease bolt with thy pease he will have,
  his houshold to feede and his hog:
Now stealeth he, now will he crave,
  and now will he coosen and cog.
In Bridewell a number be stript,
  lesse woorthie than theefe to be whipt.

The oxboy, as ill is as hee,
   or worser, if worse may be found:
For spoiling from thine and from thee,
   of grasse and of corne on the ground.
Laie never so well for to save it,
   by night or by daie he will have it.

What orchard unrobbed escapes?
   or pullet dare walke in their jet?
But homeward or outward (like apes)
   they count it their owne they can get.
Lord, if ye doo take them, what sturs!
   how hold they togither like burs!

For commons these commoners crie,
   enclosing they may not abide:
Yet some be not able to bie
   a cow with hir calfe by hir side.
Nor laie not to live by their wurke,
   but theevishlie loiter and lurke.

The Lord of the towne is to blame,
   for these and for many faults mo.
For that he doth knowe of the same,
   yet lets it unpunished go.
Such Lords ill example doth give,
   where verlets and drabs so may live.

What footpathes are made, and how brode!
   annoiance too much to be borne:
With horse and with cattle what rode
   is made thorow erie mans corne!
Where champions ruleth the roste,
   there dailie disorder is moste.

Their sheepe when they drive for to wash,
  how careles such sheepe they doo guide!
The fermer they leave in the lash,
  with losses on everie side.
Though any mans corne they doo bite,
  they will not alow him a mite.

What hunting and hauking is there!
  corne looking for sickle at hand:
Actes lawles to doo without feare,
  how yeerlie togither they band.
More harme to another to doo,
  than they would be done so untoo.

More profit is quieter found
  (where pastures in severall bee:)
Of one seelie aker of ground,
  than champion maketh of three.
Againe what a joie is it knowne,
  when men may be bold of their owne!

Champion.   The tone is commended for graine,
  yet bread made of beanes they doo eate:
Severall.   The tother for one loafe have twaine,
  of mastlin, of rie, or of wheate.
The champion liveth full bare,
  when woodland full merie doth fare.

Champion.   Tone giveth his corne in a darth,
  to horse, sheepe, and hog every daie;
Severall.   The tother give cattle warme barth,
  and feede them with strawe and with haie.
Corne spent of the tone so in vaine:
  the tother doth sell to his gaine.

Tone barefoote and ragged doth go,   *Champion.*
  and readie in winter to sterve:
When tother ye see doo not so,   *Severall.*
  but hath that is needfull to serve.
Tone paine in a cotage doth take,
  when tother trim bowers doo make.

Tone laieth for turfe and for sedge,   *Champion.*
  and hath it with woonderfull suit:
When tother in everie hedge,   *Severall.*
  hath plentie of fewell and fruit.
Evils twentie times worser than thease,
  enclosure quickly would ease.

In woodland the poore men that have   *Severall.*
  scarse fully two akers of land,
More merily live and doo save,
  than tother with twentie in hand.
Yet paie they as much for the twoo
  as tother for twentie must doo.

The labourer comming from thence,
  in woodland to worke any where:
(I warrant you) goeth not hence,
  to worke anie more again there.
If this same be true (as it is:)
  why gather they nothing of this?

The poore at enclosing doo grutch,
  because of abuses that fall,
Least some man should have but too much,
  and some againe nothing at all.
If order might therein be found,
  what were to the severall ground?

THE DESCRIPTION OF AN ENVIOUS
AND NAUGHTIE NEIGHBOUR

Chapter 53

An envious neighbour is easie to finde,
His cumbersome fetches are seldome behinde.
His hatred procureth from naughtie to wurse,
His friendship like Judas that carried the purse.
His head is a storehouse, with quarrels full fraught,
His braine is unquiet, till all come to naught.
His memorie pregnant, old evils to recite,
His mind ever fixed each evill to requite.
His mouth full of venim, his lips out of frame,
His tongue a false witnes, his friend to defame.
His eies be promooters, some trespas to spie,
His eares be as spials, alarum to crie.
His hands be as tyrants, revenging ech thing,
His feete at thine elbow, as serpent to sting.
His breast full of rancor, like Canker to freat,
His hart like a Lion, his neighbour to eat.
His gate like a sheepebiter, fleering aside,
His looke like a coxcombe, up puffed with pride.
His face made of brasse, like a vice in a game,
His jesture like Davus, whom Terence doth name.
His brag as Thersites, with elbowes abrode.
His cheekes in his furie shall swell like a tode.
His colour like ashes, his cap in his eies,
His nose in the aire, his snout in the skies.
His promise to trust to as slipprie as ice,
His credit much like to the chance of the dice.
His knowledge or skill is in prating too much,
His companie shunned, and so be all such.

His friendship is counterfait, seldome to trust,
His dooings unluckie and ever unjust.
His fetch is to flatter, to get what he can,
His purpose once gotten, a pin for thee than.

[*In the edition of 1577 the following piece is inserted here.*]

TO LIGHT A CANDELL
before the Devill

To beard thy foes shews forth thy witt,
but helpes the matter nere a whit.

MY sonne, were it not worst
   to frame thy nature so,
That as thine use is to thy friend,
   likewise to greet thy foe:
Though not for hope of good,
   yet for the feare of evill,
Thou maist find ease so proffering up
   a candell to the devill.

This knowne, the surest way
   thine enemies wrath to swage;
If thou canst currey favour thus,
   thou shalt be counted sage.
Of truth I tell no lye,
   by proofe to well I knowe,
The stubborne want of only this
   hath brought full many lowe.

And yet to speak the trouth
   the Devill is worse then naught,
That no good turne will once deserve,
   yet looketh up so haught.
Exalt him how we please,
   and give him what we can,
Yet skarcely shall we find such Devill
   a truly honest man.

But where the mighty may
   of force the weake constraine,
It shal be wysely doone to bow
   to voyd a farther payne,
Like as in tempest great,
   where wind doth beare the stroke,
Much safer stands the bowing reede
   then doth the stubborne oke.

And chiefly when of all
   thy selfe art one of those
That fortune needes, will have to dwell
   fast by the Devils nose:
Then (though against thine hart)
   thy tongue thou must so charme
That tongue may say, where ere thou come,
   the Devill doth no man harme.

For where as no revenge
   may stand a man in steede,
As good is then an humble speech,
   as otherwise to bleede.
Like as ye see by him
   that hath a shrew to wife,
As good it is to speak her faire
   as still to live in strife.

Put thou no Devill in boote
   as once did master Shorne:
Take heede as from madde bayted bull
   to keepe thee fro his horne.
And where ye see the Devill
   so bold to wrest with lawe,
Make *congé* oft, and crouch aloofe,
   but come not in his clawe.

The scholer forth of schoole
   may boldlier take his mind,
The fields have eyes, the bushes eares,
   false birds can fetch the wind.
The further from the gone
   the safer may ye skippe,
The nerer to the carters hand
   the nerer to the whippe.

The neerer to the whippe
   the sooner comes the jerke,
The sooner that poore beast is strucke
   the sooner doth he yerke.
Some loveth for to whippe,
   to see how jerkes will smart,
In wofull taking is that horse
   that nedes must drawe in cart.

Such fellow is the Devell,
   that doth even what he list,
Yet thinketh he what ere he doth
   none ought dare say, but whist.
Take therefore heed, my sonne,
   and marke full well this song,
Learne thus with craft to claw the devell,
   else live in rest not long.

A SONET
against a slanderous tongue.

Chapter 54

DOTH darnell good, among the flowrie wheat?
Doo thistles good, so thick in fallow spide?
Doo taint wormes good, that lurke where ox should eat?
Or sucking drones, in hive where bees abide?
Doo hornets good, or these same biting gnats?
Foule swelling toades, what good by them is seene?
In house well deckt, what good doth gnawing rats?
Or casting mowles, among the meadowes greene?
Doth heavie newes make glad the hart of man?
Or noisome smels, what good doth that to health?
Now once for all, what good (shew who so can?)
Doo stinging snakes, to this our Commonwealth?

No more doth good a peevish slanderous toung,
But hurts it selfe, and noies both old and young.

A SONET
upon the Authors first seven
yeeres service.

Chapter 55

SEVEN times hath Janus tane new yeere by hand,
Seven times hath blustring March blowne forth his powre:
To drive out Aprils buds, by sea and land,
For minion Maie, to deck most trim with flowre.
Seven times hath temperate Ver, like pageant plaide,
And pleasant Æstas eke hir flowers told:
Seven times Autumnes heate hath beene delaide,
With Hyems boistrous blasts, and bitter cold.
Seven times the thirteene Moones have changed hew,
Seven times the Sunne his course hath gone about:
Seven times ech bird hir nest hath built anew,
Since first time you to serve, I choosed out.

Still yours am I, though thus the time hath past,
And trust to be, as long as life shall last.

Man minded for to thrive
must wisely lay to wive.
What hap may thereby fall
here argued find ye shall.

THE AUTHOURS DIALOGUE
betweene two Bachelers, of wiving and thriving
by Affirmation and Objection.

Chapter 56

Affirmation.
FREND, where we met this other day,
We heard one make his mone and say,
   Good Lord, how might I thrive?
We heard an other answere him,
Then make thee handsome, trick and trim,
   And lay in time to wive.
Objection.
And what of that, say you to mee?
Do you your selfe thinke that to be
   The best way for to thrive?
If truth were truely bolted out,
As touching thrift, I stand in dout,
   If men were best to wive.
Affirmation.
There is no doubt, for prove I can,
I have but seldome seene that man
   Which could the way to thrive:
Untill it was his happie lot,
To stay himselfe in some good plot,
   And wisely then to wive.

Objection.
And I am of an other minde,
For by no reason can I finde,
   How that way I should thrive:
For where as now I spend a pennie,
I should not then be quit with mennie,
   Through bondage for to wive.

Affirmation.
Not so, for now where thou dost spend,
Of this and that, to no good end,
   Which hindereth thee to thrive:
Such vaine expences thou shouldst save,
And daily then lay more to have,
   As others do that wive.

Objection.
Why then do folke this proverbe put,
The blacke oxe neare trod on thy fut,
   If that way were to thrive?
Hereout a man may soone picke forth,
Few feeleth what a pennie is worth,
   Till such time as they wive.

Affirmation.
It may so chaunce as thou doest say,
This lesson therefore beare away,
   If thereby thou wilt thrive:
Looke ere thou leape, see ere thou go,
It may be for thy profite so,
   For thee to lay to wive.

Objection.
It is too much we dailie heare,
To wive and thrive both in a yeare,
   As touching now to thrive:
I know not herein what to spie,
But that there doth small profite lie,
   To fansie for to wive.

Affirmation.
In deede the first yeare oft is such,
That fondly some bestoweth much,
   A let to them to thrive:
Yet other moe may soone be founde,
Which getteth many a faire pounde,
   The same day that they wive.

Objection.
I graunt some getteth more that day,
Than they can easily beare away,
   Nowe needes then must they thrive:
What gaineth such thinke you by that?
A little burden, you wote what,
   Through fondnesse for to wive.

Affirmation.
Thou seemest blinde as mo have bin,
It is not beautie bringeth in
   The thing to make thee thrive:
In womankinde, see that ye do
Require of hir no gift but two,
   When ere ye minde to wive.

Objection.
But two, say you? I pray you than
Shew those as briefly as you can,
   If that may helpe to thrive:
I weene we must conclude anon,
Of those same twaine to want the ton,
   When ere we chance to wive.

Affirmation.

Honestie and hus-wiferie.

An honest huswife, trust to mee,
Be those same twaine, I say to thee,
   That helpe so much to thrive:
As honestie farre passeth golde,
So huswiferie in yong and olde,
   Do pleasure such as wive.

Objection.

The honestie in deede I graunt,
Is one good point the wife should haunt,
   To make hir husband thrive:
But now faine would I have you show,
How should a man good huswife know,
   If once he hap to wive?

Affirmation.

A huswife good betimes will rise,
And order things in comelie wise,
   Hir minde is set to thrive:
Upon hir distaffe she will spinne,
And with hir needle she will winne,
   If such ye hap to wive.

Objection.

It is not idle going about,
Nor all day pricking on a clout,
   Can make a man to thrive:
Or if there be no other winning,
But that the wife gets by hir spinning,
   Small thrift it is to wive.

Affirmation.

Some more than this yet do shee shall,
Although thy stocke be verie small,
   Yet will shee helpe thee thrive:
Lay thou to save, as well as she,
And then thou shalt enriched be,
   When such thou hapst to wive.

Objection.

If she were mine, I tell thee troth,
Too much to trouble hir I were loth,
   For greedines to thrive:
Least some should talke, as is the speech,
The good wives husband weares no breech,
   If such I hap to wive.

Affirmation.
What hurts it thee what some do say,
If honestlie she take the way
   To helpe thee for to thrive?
For honestie will make hir prest,
To doo the thing that shall be best,
   If such ye hap to wive.

Objection.
Why did *Diogenes* say than,
To one that askt of him time whan,
   Were best to wive to thrive?
Not yet (quoth he) if thou be yong,
If thou waxe old, then holde thy tong,
   It is too late to wive.

Affirmation.
Belike he knew some shrewish wife,
Which with hir husband made such strife,
   That hindered him to thrive:
Who then may blame him for that clause,
Though then he spake as some had cause,
   As touching for to wive?

Objection.
Why then I see to take a shrew,
(As seldome other there be few)
   Is not the way to thrive:
So hard a thing I spie it is,
The good to chuse, the shrew to mis,
   That feareth me to wive.

Affirmation.
She may in something seeme a shrew,
Yet such a huswife as but few,
   To helpe thee for to thrive:
This proverbe looke in mind ye keepe,
As good a shrew is as a sheepe,
   For you to take to wive.

Objection.

Now be she lambe or be she eaw,
Give me the sheepe, take thou the shreaw,
   See which of us shall thrive:
If she be shrewish thinke for troth,
For all her thrift I would be loth
   To match with such to wive.

Affirmation.

Tush, farewell then, I leave you off,
Such fooles as you that love to scoff,
   Shall seldome wive to thrive:
Contrarie hir, as you do me,
And then ye shall, I warrant ye,
   Repent ye if ye wive.

Objection.

Friend, let us both give justly place,
To wedded man to judge this cace,
   Which best way is to thrive:
For both our talke as seemeth plaine,
Is but as hapneth in our braine,
   To will or not to wive.

*Wedded mans judgement
Upon the former argument.*

As Cock that wants his mate, goes roving al about,   *Moderator.*
With crowing early and late, to find his lover out:
And as poore sillie hen, long wanting cock to guide,
Soone droopes and shortly then beginnes to
   peake aside:
Even so it is with man and wife, where govern-
   ment is found,
The want of ton the others life doth shortly soone
   confound.

In jest and in earnest, here argued ye finde,
That husband and huswife togither must dwell,
And thereto the judgement of wedded mans minde,
That husbandrie otherwise speedeth not well:
So somewhat more nowe I intende for to tell,
Of huswiferie like as of husbandrie tolde,
How huswifelie huswife helpes bring in the golde.

> *Thus endeth the booke of*
> Husbandrie.

THE POINTS OF HUSWIFERIE
UNITED TO THE COMFORT OF HUSBANDRIE
newly corrected and amplified, with divers good lessons
for housholders to recreate the Reader,
as by the Table at the end hereof
more plainlie may appeere.

Set forth by
THOMAS TUSSER
Gentleman.

TO THE RIGHT HONORABLE
and my especiall good Ladie and Maistres,
THE LADIE PAGET

Though danger be mickle,
and favour so fickle,
Yet dutie doth tickle
   my fansie to wright:
Concerning how prettie,
how fine and how nettie,
Good huswife should jettie,
   from morning to night.

Not minding by writing,
to kindle a spiting,
But shew by enditing,
   as afterward told:
How husbandrie easeth,
to huswiferie pleaseth,
And manie purse greaseth
   with silver and gold.

For husbandrie weepeth,
where huswiferie sleepeth,
And hardly he creepeth,
   up ladder to thrift:
That wanteth to bold him,
thrifts ladder to hold him,
Before it be told him,
   he falles without shift.

Least many should feare me,
and others forsweare me,
Of troth I doo beare me
   upright as ye see:
Full minded to loove all,
and not to reproove all,
But onely to moove all,
   good huswives to bee.

For if I should mind some,
or descant behind some,
And missing to find some,
   displease so I mought:
Or if I should blend them,
and so to offend them,
What stur I should send them
   I stand in a dought.

Though harmles ye make it
and some doo well take it,
If others forsake it,
   what pleasure were that?
Naught else but to paine me,
and nothing to gaine me,
But make them disdaine me
   I wot ner for what.

Least some make a triall,
as clocke by the diall,
Some stand to deniall,
   some murmur and grudge:
Give judgement I pray you,
for justlie so may you,
So fansie, so say you,
   I make you my judge.

In time, ye shall try me,
by troth, ye shall spy me,
So finde, so set by me,
   according to skill:
How ever tree groweth,
the fruit the tree showeth,
Your Ladiship knoweth,
   my hart and good will.

Thogh fortune doth measure,
and I doo lacke treasure,
Yet if I may pleasure
   your Honour with this:
Then will me to mend it,
or mend er ye send it,
Or any where lend it,
   if ought be amis.

               Your Ladiships Servant,
                  THOMAS TUSSER.

TO THE READER

NOW listen, good huswives, what dooings are here
   set foorth for a daie, as it should for a yere.
Both easie to follow, and soone to atchive,
   for such as by huswiferie looketh to thrive.

The forenoone affaires, till dinner (with some,)
   then after noone dooings, till supper time come.
With breakfast and dinner time, sup, and to bed,
   standes orderlie placed, to quiet thine hed.

The meaning is this, for a daie what ye see,
   that monthlie and yeerlie continued must bee.
And hereby to gather (as proove I intend),
   that huswivelie matters have never an end.

I have not, by heare say, nor reading in booke,
   set out (peradventure) that some cannot brooke,
Nor yet of a spite, to be dooing with enie,
   but such as have skared me many a penie.

If widow, both huswife and husband may be,
   what cause hath a widower lesser than she?
Tis needfull that both of them looke well about:
   too careles within, and too lasie without.

Now therefore, if well ye consider of this,
   what losses and crosses comes dailie amis.
Then beare with a widowers pen as ye may:
   though husband of huswiferie somewhat doth say.

THE PREFACE
to the booke of Huswiferie

TAKE weapon away, of what force is a man?
Take huswife from husband, and what is he than?

As lovers desireth together to dwell,
So husbandrie loveth good huswiferie well.

Though husbandrie seemeth to bring in the gaines,
Yet huswiferie labours seeme equall in paines.

Some respit to husbands the weather may send,
But huswives affaires have never an end.

>   As true as thy faith,
>   Thus huswiferie saith.

I SERVE for a daie, for a weeke, for a yere,   The praise
For life time, for ever, while man dwelleth here.   of hus-
For richer, for poorer, from North to the South,   wiferie.
For honest, for hardhead, or daintie of mouth.
For wed and unwedded, in sicknes and health,
For all that well liveth, in good Commonwealth.
For citie, for countrie, for Court, and for cart,
To quiet the head, and to comfort the hart.

A DESCRIPTION
of Huswife and Huswiferie

OF huswife doth huswiferie challenge that name,
   of huswiferie huswife doth likewise the same,
Where husband and husbandrie joineth with thease,
   there wealthines gotten is holden with ease.

The name of a huswife what is it to say?
   the wife of the house, to the husband a stay.
If huswife doth that, as belongeth to hur:
   if husband be godlie, there needeth no stur.

The huswife is she that to labour doth fall,
   the labour of hir I doo huswiferie call.
If thrift by that labour be honestlie got:
   then is it good huswiferie, else is it not.

The woman the name of a huswife doth win,
   by keeping hir house, and of dooings therein.
And she that with husband will quietly dwell,
   must thinke on this lesson, and follow it well.

INSTRUCTIONS TO HUSWIFERIE

Serve God is the furst,
True love is not wurst.

A DAILIE good lesson, of huswife in deede,
   is God to remember, the better to speede.

An other good lesson, of huswiferie thought,
   is huswife with husband to live as she ought.

Wife comely no griefe,
Man out, huswife chiefe.

Though trickly to see to, be gallant to wive,
   yet comely and wise is the huswife to thrive.

When husband is absent, let huswife be chiefe,
   and looke to their labour that eateth hir biefe.

Both out not allow,
Keepe house huswife thow.

Where husband and huswife be both out of place,
   there servants doo loiter, and reason their cace.

The huswife so named (of keeping the house,)
   must tend on hir profit, as cat on the mouse.

Seeke home for rest,
For home is best.

As huswives keepe home, and be stirrers about,
   so speedeth their winnings, the yeere thorow out.

Though home be but homely, yet huswife is taught,
   that home hath no fellow to such as have aught.

Use all with skill,
Aske what ye will.

Good usage with knowledge, and quiet withall,
   make huswife to shine, as the sunne on the wall.

What husband refuseth all comely to have,
   that hath a good huswife, all willing to save.

> *Be readie at neede,*
> *All thine to feede.*

The case of good huswives, thus daily doth stand,
   what ever shall chance, to be readie at hand.

This care hath a huswife all daie in hir hed,
   that all thing in season be huswifelie fed.

> *By practise go muse,*
> *How houshold to use.*

Dame practise is she that to huswife doth tell,
   which way for to governe hir familie well.

Use labourers gently, keepe this as a lawe,
   make childe to be civill, keepe servant in awe.

> *Who careles doe live,*
> *Occasion doe give.*

Have everie where a respect to thy waies,
   that none of thy life any slander may raies.

What many doo knowe, though a time it be hid,
   at length will abrode, when a mischiefe shall bid.

> *No neighbour reproove,*
> *Doe so to have loove.*

The love of thy neighbour shall stand thee in steede,
   the poorer, the gladder, to helpe at a neede.

Use friendly thy neighbour, else trust him in this,
   as he hath thy friendship, so trust unto his.

> *Strike nothing unknowne,*
> *Take heede to thine owne.*

Revenge not thy wrath upon any mans beast,
   least thine by like malice be bid to like feast.

What husband provideth with monie his drudge,
   the huswife must looke to, which waie it doth trudge.

A Digression

NOW, out of the matter, this lesson I ad,
   concerning cock crowing, what profit is had.
Experience teacheth, as true as a clock:
   how winter night passeth, by marking the cock.

Cock croweth at midnight, times few above six,
   with pause to his neighbour, to answere betwix.
At three a clock thicker, and then as ye knowe,
   like all in to Mattens, neere daie they doo crowe.

At midnight, at three and an hower ere day,   *Cocke*
   they utter their language, as well as they may.   *crowing*.
Which who so regardeth what counsell they give,
   will better love crowing, as long as they live.

> *For being afraid,*
> *Take heede good maid:*
> *Marke crowing of cock,*
> *For feare of a knock.*

> *The first cock croweth.*

Ho, Dame it is midnight: what rumbling is that?

> *The next cock croweth.*

Take heede to false harlots, and more, ye wot what.

> *If noise ye heare,*
> *Looke all be cleare:*
> *Least drabs doe noie thee,*
> *And theeves destroie thee.*

> *The first cock croweth.*

Maides, three a clock, knede, lay your bucks, or go brew,

> *The next cock croweth.*

And cobble and botch, ye that cannot buie new.

> *Till cock crow agen,*
> *Both maidens and men:*
> *Amend now with speede,*
> *That mending doth neede.*

> *The first cock croweth.*

Past five a clock, Holla: maid, sleeping beware,

> *The next cock croweth.*

Least quickly your Mistres uncover your bare.

> *Maides, up I beseech yee,*
> *Least Mistres doe breech yee:*
> *To worke and away,*
> *As fast as ye may.*

HUSWIFERIE

MORNING WORKES

No sooner some up,
But nose is in cup.

GET up in the morning as soone as thou wilt,
with overlong slugging good servant is spilt.

Some slovens from sleeping no sooner get up,
but hand is in aumbrie, and nose in the cup.

That early is donne,
Count huswifely wonne.

Some worke in the morning may trimly be donne,   *Morning*
that all the day after can hardly be wonne.   *workes.*

Good husband without it is needfull there be,
good huswife within as needfull as he.

Cast dust into yard,
And spin and go card.

Sluts corners avoided shall further thy health,
much time about trifles shall hinder thy wealth.

Set some to peele hempe or else rishes to twine,
to spin and to card, or to seething of brine.

Grind mault for drinke,
See meate do not stinke.

Set some about cattle, some pasture to vewe,
some mault to be grinding against ye do brewe.

Some corneth, some brineth, some will not be taught,
where meate is attainted, there cookrie is naught.

BREAKEFAST DOINGS

> *To breakefast that come,*
> *Give erie one some.*

Breakefast CALL servants to breakefast by day starre appere,
a snatch and to worke, fellowes tarrie not here.

Let huswife be carver, let pottage be heate,
a messe to eche one, with a morsell of meate.

> *No more tittle tattle,*
> *Go serve your cattle.*

What tacke in a pudding, saith greedie gut wringer,
give such ye wote what, ere a pudding he finger.

Let servants once served, thy cattle go serve,
least often ill serving make cattle to sterve.

HUSWIFELY ADMONITIONS

> *Learne you that will thee,*
> *This lesson of mee.*

Thee for thrive. NO breakefast of custome provide for to save,
but onely for such as deserveth to have.

No shewing of servant what vittles in store,
shew servant his labour, and shew him no more.

> *Of havocke beware,*
> *Cat nothing will spare.*

Where all thing is common, what needeth a hutch?
where wanteth a saver, there havocke is mutch.

Where window is open, cat maketh a fray,
yet wilde cat with two legs is worse by my fay.

Looke well unto thine,
Slut slouthfull must whine.

An eie in a corner who useth to have,
revealeth a drab, and preventeth a knave.

Make maide to be clenly, or make hir crie creake,
and teach hir to stirre, when hir mistresse doth speake.

Let hollie wand threate,
Let fisgig be beate.

A wand in thy hand, though ye fight not at all,
makes youth to their businesse better to fall.

For feare of foole had I wist cause thee to waile,
let fisgig be taught to shut doore after taile.

Too easie the wicket,
Will still appease clicket.

With hir that will clicket make daunger to cope,
least quickly hir wicket seeme easie to ope.

As rod little mendeth where maners be spilt,
so naught will be naught say and do what thou wilt.

Fight seldome ye shall
But use not to brall.

Much bralling with servant, what man can abide?
pay home when thou fightest, but love not to chide.

As order is heavenly where quiet is had,
so error is hell, or a mischiefe as bad.

What better a lawe.
Than subjects in awe?

Such awe as a warning will cause to beware,
doth make the whole houshold the better to fare.

The lesse of thy counsell thy servants doe knowe,
Their dutie the better such servants shall showe.

> *Good musicke regard,*
> *Good servants reward.*

Such servants are oftenest painfull and good,
that sing in their labour, as birdes in thee wood.

Good servants hope justly some friendship to feele,
and looke to have favour what time they do weele.

> *By once or twise*
> *Tis time to be wise.*

Take runagate Robin, to pitie his neede,
and looke to be filched, as sure as thy creede.

Take warning by once, that a worse do not hap,
foresight is the stopper of many a gap.

> *Some change for a shift,*
> *Oft change, small thrift.*

Make fewe of thy counsell to change for the best,
least one that is trudging infecteth the rest.

The stone that is rolling can gather no mosse,
for maister and servant, oft changing is losse.

> *Both liberall sticketh,*
> *Some provender pricketh.*

One liberall. One dog for a hog, and one cat for a mouse,
one readie to give is ynough in a house:

One gift ill accepted, keepe next in thy purse,
whom provender pricketh are often the wurse.

BREWING

Brew somewhat for thine,
Else bring up no swine.

WHERE brewing is needfull, be brewer thy selfe,   *Brewing.*
what filleth the roofe will helpe furnish the shelfe:

In buieng of drinke, by the firkin or pot,
the tallie ariseth, but hog amendes not.

Well brewed, worth cost,
Ill used, halfe lost.

One bushell well brewed, outlasteth some twaine,
and saveth both mault, and expences in vaine.

Too new is no profite, too stale is as bad,
drinke dead or else sower makes laborer sad.

Remember good Gill,
Take paine with thy swill.

Seeth grains in more water, while grains be yet hot,   *Seething*
and stirre them in copper, as poredge in pot.   *of graines.*

Such heating with straw, to have offall good store,
both pleaseth and easeth, what would ye have more?

BAKING

Newe bread is a drivell.
Much crust is as evill.

NEW bread is a waster, but mouldie is wurse,   *Baking.*
what that way dog catcheth, that loseth the purse.

Much dowebake I praise not, much crust is as ill,
the meane is the Huswife, say nay if ye will.

COOKERIE

Good cookerie craveth,
Good turnebroch saveth.

Cookerie. GOOD cooke to dresse dinner, to bake and to brewe,
deserves a rewarde, being honest and trewe.

Good diligent turnebroch and trustie withall,
is sometime as needfull as some in the hall.

DAIRIE

Good dairie doth pleasure,
Ill dairie spendes treasure.

Dairie. GOOD huswife in dairie, that needes not be tolde,
deserveth hir fee to be paid hir in golde.

Ill servant neglecting what huswiferie saies,
deserveth hir fee to be paid hir with baies.

Good droie woorth much.
Marke sluts and such.

Good droie to serve hog, to helpe wash, and to milke,
more needfull is truelie than some in their silke.

Though homelie be milker, let cleanlie be cooke,
for a slut and a sloven be knowne by their looke.

In dairie no cat,
Laie bane for a rat.

Traps for rats. Though cat (a good mouser) doth dwell in a house,
yet ever in dairie have trap for a mouse.

Take heede how thou laiest the bane for the rats,
for poisoning servant, thy selfe and thy brats.

SCOURING

*No scouring for pride,
Spare kettle whole side.*

THOUGH scouring be needfull, yet scouring too mutch,   *Scouring.*
is pride without profit, and robbeth thine hutch.

Keepe kettles from knocks, set tubs out of Sun,
for mending is costlie, and crackt is soone dun.

WASHING

*Take heede when ye wash,
Else run in the lash.*

MAIDS, wash well and wring well, but beat ye wot how,   *Washing.*
if any lack beating, I feare it be yow.

In washing by hand, have an eie to thy boll,
for launders and millers, be quick of their toll.

*Drie sunne, drie winde,
Safe binde, safe finde.*

Go wash well, saith Sommer, with sunne I shall drie,
go wring well, saith Winter, with winde so shall I.

To trust without heede is to venter a joint,
give tale and take count, is a huswifelie point.

*Where many be packing,
Are manie things lacking.*

Where hens fall a cackling, take heede to their nest,
where drabs fall a whispring, take heede to the rest.

Through negligent huswifes, are many things lacking,
and Gillet suspected will quickly be packing.

MALTING

*Ill malting is theft,
Wood dride hath a weft.*

Malting.   HOUSE may be so handsome, and skilfulnes such,
to make thy owne malt, it shall profit thee much.

Som drieth with strawe, and some drieth with wood,
wood asketh more charge, and nothing so good.

*Take heede to the kell,
Sing out as a bell.*

Be suer no chances to fier can drawe,
the wood, or the furzen, the brake or the strawe.

Let Gillet be singing, it doth verie well,
to keepe hir from sleeping and burning the kell.

*Best dride best speedes,
Ill kept, bowd breedes.*

Malt being well speered, the more it will cast,
malt being well dried, the longer will last.

Long kept in ill soller, (undoubted thou shalt.)
through bowds without number loose quickly thy malt.

DINNER MATTERS

*For hunger or thirst,
Serve cattle well first.*

Dinner time.   BY noone see your dinner, be readie and neate,
let meate tarrie servant, not servant his meate.

Plough cattle a baiting, call servant to dinner,
the thicker togither, the charges the thinner.

Togither is best,
For hostis and gest.

Due season is best, altogither is gay,
dispatch hath no fellow, make short and away.

Beware of Gill laggoose, disordring thy house,
mo dainties who catcheth, than craftie fed mouse?

Let such have ynough,
That follow the plough.

Give servant no dainties, but give him ynough,
too many chaps walking, do begger the plough.

Poore seggons halfe starved worke faintly and dull,
and lubbers doo loiter, their bellies too full.

Give never too much,
To lazie and such.

Feede lazie that thresheth a flap and a tap,
like slothfull, that all day be stopping a gap.

Some litherly lubber more eateth than twoo,
yet leaveth undone that another will doo.

Where nothing will last,
Spare such as thou hast.

Some cutteth thy linnen, some spoileth* their broth,
bare table to some doth as well as a cloth.

Treene dishes be homely, and yet not to lack,
where stone is no laster take tankard and jack.

Knap boy on the thums,
And save him his crums.

That pewter is never for manerly feastes,
that daily doth serve so unmanerly beastes.

Some gnaweth and leaveth, some crusts and some crums.
eat such their own levings, or gnaw their own thums.

**spilleth.*   1577.

> *Serve God ever furst,*
> *Take nothing at wurst.*

Grace before   At dinner, at supper, at morning, at night,
& after meate.   give thankes unto God, for his gifts so in sight.

Good husband and huswife, will sometime alone,
make shift with a morsell and picke of a bone.

> *Inough thou art tolde,*
> *Too much will not holde.*

Three dishes well dressed, and welcome withall,
both pleaseth thy friend and becommeth thine hall.

Enough is a plentie, too much is a pride,
the plough with ill holding, goes quicklie aside.

AFTER NOONE WORKES

> *Make companie breake,*
> *Go cherish the weake.*

After noone   WHEN dinner is ended, set servants to wurke,
workes.   and follow such fellowes as loveth to lurke.

To servant in sicknesse see nothing ye grutch,
a thing of a trifle shall comfort him mutch.

> *Who manie do feede,*
> *Save much they had neede.*

Put chippings in dippings, use parings to save,
fat capons or chickens that lookest to have.

Save droppings and skimmings, how ever ye doo,
for medcine for cattell, for cart and for shoo.

> *Leane capon unmeete,*
> *Deere fed is unsweete.*

Such ofcorne as commeth give wife to hir fee,
feede willingly such as do helpe to feede thee.

Though fat fed is daintie, yet this I thee warne,
be cunning in fatting for robbing thy barne.

> *Peece hole to defende.*
> *Things timely amende.*

Good semsters be sowing of fine pretie knackes,
good huswifes be mending and peecing their sackes.

Though making and mending be huswifely waies,
yet mending in time is the huswife to praies.

> *Buie newe as is meete,*
> *Marke blanket and sheete.*

Though Ladies may rend and buie new ery day,
good huswifes must mend and buie new as they may.

Call quarterly servants to court and to leete,
write everie Coverlet, Blanket, and Sheete.

> *Shift slovenly elfe,*
> *Be gayler thy selfe.*

Though shifting too oft be a theefe in a house,
yet shift slut and sloven for feare of a louse.

Graunt doubtfull no key of his chamber in purse,
least chamber doore lockt be to theeverie a nurse.

> *Save feathers for gest,*
> *These other rob chest.*

Save wing for a thresher, when Gander doth die,   *Save*
save fether of all thing, the softer to lie.   *feathers.*

Much spice is a theefe, so is candle and fier,
sweete sauce is as craftie as ever was frier.

> *Wife make thine owne candle,*
> *Spare pennie to handle.*

Candle making. Provide for thy tallow, ere frost commeth in,
and make thine owne candle, ere winter begin.

If pennie for all thing be suffred to trudge,
trust long, not to pennie, to have him thy drudge.

EVENING WORKES

> *Time drawing to night,*
> *See all things go right.*

Evening workes. WHEN hennes go to roost go in hand to dresse meate,
serve hogs and to milking and some to serve neate.

Where twaine be ynow, be not served with three,
more knaves in a companie worser they bee.

> *Make lackey to trudge,*
> *Make servant thy drudge.*

For everie trifle leave janting thy nag,
but rather make lackey of Jack boie thy wag.

Make servant at night lug in wood or a log,
let none come in emptie but slut and thy dog.

> *False knave readie prest,*
> *All safe is the best.*

Where pullen use nightly to pearch in the yard,
there two legged foxes keepe watches and ward.

See cattle well served, without and within,
and all thing at quiet ere supper begin.

> *Take heede it is needeful,*
> *True pittie is meedeful.*

No clothes in garden, no trinkets without,
no doore leave unbolted, for feare of a dout.

Thou woman whom pitie becommeth the best,
graunt all that hath laboured time to take rest.

SUPPER MATTERS

*Use mirth and good woorde,
At bed and at boorde.*

PROVIDE for thy husband, to make him good cheere, *Supper time*
make merrie togither, while time ye be heere. *huswiferie.*

At bed and at boord, howsoever befall,
what ever God sendeth be merrie withall.

*No brawling make,
No jelousie take.*

No taunts before servants, for hindring of fame,
no jarring too loude for avoyding of shame.

As fransie and heresie roveth togither,
so jealousie leadeth a foole ye wot whither.

*Tend such as ye have,
Stop talkative knave.*

Yong children and chickens would ever be eating,
good servants looke dulie for gentle intreating.

No servant at table use sausly to talke,
least tongue set at large out of measure do walke.

*No snatching at all,
Sirs, hearken now all.*

No lurching, no snatching, no striving at all,
least one go without and another have all.

Declare after Supper, take heede thereunto,
what worke in the morning ech servant shall do.

AFTER SUPPER MATTERS

Thy soule hath a clog,
Forget not thy dog.

Workes after supper. REMEMBER those children whose parents be poore, which hunger, yet dare not crave at thy doore.

Thy Bandog that serveth for diverse mishaps,
forget not to give him thy bones and thy scraps.

Make keies to be keepers,
To bed ye sleepers.

Where mouthes be many, to spend that thou hast,
set keies to be keepers, for spending too fast.

To bed after supper let drousie go sleepe,
least knave in the darke to his marrow do creepe.

Keepe keies as thy life,
Feare candle good wife.

Such keies lay up safe, ere ye take ye to rest,
of dairie, of buttrie, of cubboord and chest.

Feare candle in hailoft, in barne, and in shed,
feare flea smocke & mendbreech, for burning their bed.

See doore lockt fast,
Two keies make wast.

A doore without locke is a baite for a knave,
a locke without key is a foole that will have.

One key to two locks, if it breake is a greefe,
two keies to one locke in the ende is a theefe.

Night workes troubles hed,
Locke doores and to bed.

The day willeth done whatsoever ye bid,
the night is a theefe, if ye take not good hid.

Wash dishes, lay leavens, save fire and away,
locke doores and to bed, a good huswife will say.

> *To bed know thy guise,*
> *To rise do likewise.*

In winter at nine, and in sommer at ten,   *Bed time.*
to bed after supper both maidens and men.

In winter at five a clocke, servant arise,   *Time to rise.*
in sommer at foure is verie good guise.

> *Love so as ye may*
> *Love many a day.*

Be lowly not sollen, if ought go amisse,
what wresting may loose thee, that winne with a kisse.

Both beare and forebeare now and then as ye may,
then, wench God a mercie, thy husband will say.

THE PLOUGHMANS FEASTING DAIES

> *This would not be slept,*
> *Old guise must be kept.*

GOOD huswives, whom God hath enriched ynough,
   forget not the feastes that belong to the plough.

The meaning is onelie to joie and be glad,
   for comfort with labour is fit to be had.

Plough Monday.

Plough Monday, next after that Twelftide is past,   *Leicestershire.*
   bids out with the plough, the woorst husband is last.

If ploughman get hatchet or whip to the skreene,
   maides loseth their cock if no water be seene.

Shroftide.

At Shroftide to shroving, go thresh the fat hen,   *Essex and*
   if blindfild can kill hir, then give it thy men.   *Suffolke.*

Maides, fritters and pancakes ynow see ye make:
   let slut have one pancake, for companie sake.

Sheepe shearing.

*Northamp-
tonshire.*

Wife make us a dinner, spare flesh neither corne,
  make wafers and cakes, for our sheepe must be shorne.

At sheepe shearing neighbours none other thing crave,
  but good cheere and welcome like neighbours to have.

The wake day.

Leicestershire

Fill oven full of flawnes, Ginnie passe not for sleepe,
  to morow thy father his wake day will keepe.

Then everie wanton may daunce at hir will,
  both Tomkin with Tomlin, and Jankin with Gill.

Harvest home.

For all this good feasting, yet art thou not loose,
  till ploughman thou givest his harvest home goose.

Though goose go in stubble, I passe not for that,
  let goose have a goose, be she leane, be she fat.

Seede cake.

*Essex and
Suffolke.*

Wife, some time this weeke, if the wether hold cleere,
  an end of wheat sowing we make for this yeere.

Remember you therefore though I doo it not:
  the seede Cake, the Pasties, and Furmentie pot.

Twise a week roast.

Good ploughmen looke weekly, of custome and right,
  for roast meat on Sundaies and Thursdaies at night.

This dooing and keeping such custome and guise,
  they call thee good huswife, they love thee likewise.

The good huswifelie
PHYSICKE

GOOD huswives provides, ere an sicknes doo come,
   of sundrie good things in hir house to have some.
Good Aqua composita, Vineger tart,
   Rose water and treakle, to comfort the hart.

Cold herbes in hir garden for agues that burne,
   that over strong heat to good temper may turne.
While Endive and Suckerie, with Spinnage ynough,
   all such with good pot herbes should follow the plough.

Get water of Fumentorie, Liver to coole,
   and others the like, or els lie like a foole.
Conserve of the Barberie, Quinces and such,
   with Sirops that easeth the sickly so much.

Aske *Medicus* counsell, ere medcine ye make,   *Physition.*
   and honour that man, for necessities sake.
Though thousands hate physick, because of the cost,
   yet thousands it helpeth, that else should be lost.

Good broth and good keeping doo much now and than,   *Good diet.*
   good diet with wisedome best comforteth man.
In health to be stirring shall profit thee best,
   in sicknes hate trouble, seeke quiet and rest.

Remember thy soule, let no fansie prevaile,   *Thinke on*
   make readie to Godward, let faith never quaile.   *thy soule*
The sooner thy selfe thou submittest to God,   *and have a*
   the sooner he ceaseth to scourge with his rod.   *good hope.*

The good motherlie
NURSERIE

GOOD huswives take paine, and doo count it good luck,
   to make their owne brest their owne childe to give suck.
Though wrauling and rocking be noisome so neare,
   yet lost by ill nursing is woorser to heare.

But one thing I warne thee, let huswife be nurse,
   least husband doo find thee too franke with his purse.
What hilback and filbellie maketh away,
   that helpe to make good, or else looke for a fraie.

Give childe that is fitly, give babie the big,
   give hardnes to youth and to roperipe a twig.
Wee find it not spoken so often for naught,
   that children were better unborne than untaught.

Some cockneies with cocking are made verie fooles,
   fit neither for prentise, for plough, nor for schooles.
Teach childe to aske blessing, serve God, and to church,
   then blesse as a mother, else blesse him with burch.
Thou huswife thus dooing, what further shall neede?
   but all men to call thee good mother in deede.

Thinke on the poore

REMEMBER the poore, that for Gods sake doo call,
   for God both rewardeth and blesseth withall.
Take this in good part, whatsoever thou bee:
   and wish me no woorse than I wish unto thee.

A COMPARISON
betweene good huswiferie and evill

Comparing togither, good huswife with bad,
The knowledge of either, the better is had.

Il huswiferie lieth
   till nine of the clock.
Good huswiferie trieth
   to rise with the cock.

Ill huswiferie tooteth,
   to make hir selfe brave.
Good huswiferie looketh
   what houshold must have.

Ill huswiferie trusteth
   to him and to hir.
Good huswiferie lusteth
   hir selfe for to stir.

Ill huswiferie careth
   for this nor for that.
Good huswiferie spareth
   for feare ye wot what.

Ill huswiferie pricketh
   hir selfe up in pride.
Good huswiferie tricketh
   hir house as a bride.

Ill huswiferie othing
   or other must crave.
Good huswiferie nothing.
   but needfull will have.

Ill huswiferie mooveth
   with gossep to spend.
Good huswiferie loveth
   hir houshold to tend.

Il huswiferie wanteth
   with spending too fast.
Good huswiferie scanteth
   the lenger to last.

Ill huswiferie easeth
   hir selfe with unknowne.
Good huswiferie pleaseth
   hir selfe with hir owne.

Il huswiferie brooketh
   mad toies in hir hed.
Good huswiferie looketh
   that all things be fed.

Il huswifrie bringeth
   a shilling to naught.
Good huswiferie singeth,
   hir cofers full fraught.

Il huswiferie rendeth,
   and casteth aside.
Good huswiferie mendeth,
   else would it go wide.

Il huswifrie sweepeth
  her linnen to gage.
Good huswiferie keepeth,
  to serve hir in age.

Ill huswiferie pineth,
  not having to eate.
Good huswiferie dineth,
  with plentie of meate.

Il huswiferie craveth
  in secret to borow.
Good huswiferie saveth
  to day for to morow.

Ill huswiferie letteth
  the Divell take all.
Good huswiferie setteth
  good brag of a small.

Good huswife good fame hath of best in the towne,
Ill huswife ill name hath of everie clowne.

*Thus endeth the booke of
Huswiferie*

For men a perfect warning
How childe shall come by larning.

ALL you that faine would learne the perfect waie,
To have your childe in Musick something seene,
Aske nature first what thereto she doth saie,
Ere further suite ye make to such a Queene.
For doubtlesse *Grossum caput* is not he
Of whom the learned Muses seene will be.

Once tride that nature trim hath done hir part,
And Ladie Musick farre in love withall,
Be wise who first doth teach thy childe that Art,
Least homelie breaker mar fine ambling ball.
Not rod in mad braines hand is that can helpe,
But gentle skill doth make the proper whelpe.

Where choise is hard, count good for well a fine,
Skill mixt with will, is he that teacheth best:
Let this suffice for teaching childe of thine,
Choose quickly well for all the lingring rest.
Mistaught at first how seldome prooveth well?
Trim taught, O God, how shortly doth excell?

Although as ships must tarrie winde and tide,
And perfect howers abide their stinted time;
So likewise, though of learning dailie tride,
Space must be had ere wit may thereto clime.
Yet easie steps, and perfect way to trust,
Doth cause good speede, confesse of force we must.

Thus in the childe though wit ynough we finde,
And teacher good neere hand or other where,
And time as apt as may be thought with minde,
Nor cause in such thing much to doubt or feare.
Yet cocking Mams, and shifting Dads from schooles,
Make pregnant wits to proove unlearned fooles.

Ere learning come, to have first art thou taught,
Apt learning childe, apt time that thing to frame,
Apt cunning man to teach, else all is naught,
Apt parents, glad to bring to passe the same.
On such apt ground the Muses love to bilde,
This lesson learne; adue else learned child.

The description of a womans age
by vi times xiiij yeeres prentiship,
with a lesson to the same.

14 Two first seven yeeres, for a rod they doe whine,
28 Two next, as a perle in the world they doe shine,
42 Two next, trim beautie beginneth to swerve,
56 Two next, for matrones or drudges they serve,
70 Two next, doth crave a staffe for a stay,
84 Two next, a beere to fetch them away.

A LESSON { Then purchase some pelfe, / by fiftie and three: / or buckle thy selfe, / a drudge for to bee.

THE INHOLDERS POSIE

AT meales my friend who vitleth here, and sitteth with his host,
Shall both be sure of better chere, and scape with lesser cost.

But he that will attendance have, a chamber by himselfe,
Must more regard what pains do crave than passe of wordly pelfe.

Let no man looke to purchase linne with pinching by the waie,
But laie before he takes his Inne to make his purse to paie.

For nothing paie and nothing praie, in Inne it is the gise,
Where no point gain, there no point pain, think this if you be wise.

For toiling much and spoiling more, great charge smal gains or none,
Soone sets thine host at needams shore, to crave the beggers bone.

Foreseeing this, come day or night, take up what place ye please.
Use mine as thine, let fortune spight, and boldly take thine ease.

CERTAINE TABLE LESSONS

FRIEND, eat lesse, and drinke lesse, and buie thee a knife,
   else looke for a carver not alway too rife.
Some kniveles their daggers for braverie weare,
   that often for surfetting neede not to feare.

At dinner and supper the table doth crave
   good fellowly neighbour good manner to have.
Advise thee well therefore, ere tongue be too free,
   or slapsauce be noted too saucie to bee.

If anything wanteth or seemeth amis,
   to call for or shew it, good maner it is.
But busie fault finder, and saucie withall,
   is roister like ruffen, no manner at all.

Some cutteth the napkin, some trencher will nick,
   some sheweth like follie, in many a trick.
Let such apish bodie so toieng at meate,
   go toie with his nodie, like ape in the streate.

Some commeth unsent for, not for thy good cheere,
   but sent as a spiall, to listen and heere.
Which being once knowne, for a knave let him go,
   for knave will be knavish, his nature is so.

LESSONS FOR WAITING SERVANTS

ONE diligent serviture, skilfull to waight,
   more comelieth thy table than other some eight,
That stand for to listen, or gasing about,
   not minding their dutie, within nor without.
Such waiter is fautie that standeth so by,
   unmindful of service, forgetting his ey.
If maister to such give a bone for to gnaw,
   he doth but his office, to teach such a daw.

Such serviture also deserveth a check,
　　that runneth out fisging with meat in his beck.
Such ravening puttocks for vittles so trim,
　　would have a good maister to puttock with him.

Who daily can suffer, or else can afoord,
　　his meat so up snatched that comes from his boord?
So tossed with cormorants, here and there some,
　　and others to want it that orderlie come?

Good serviture waieth (once dinner begon,)
　　what asketh attendance and what to be don.
So purchasing maister a praise with the best,
　　gets praise to himselfe, both of maister and gest.

HUSBANDLY POSIES FOR THE HALL

FRIEND, here I dwell, and here I have a little worldly pelfe,
Which on my friend I keepe to spend, as well as on my selfe.

What ever fare you hap to finde, take welcome for the best,
That having then disdaine thou not, for wanting of the rest.

Backbiting talk that flattering blabs know wily how to blenge,
The wise doth note, the friend doth hate, the enimie will revenge.

The wise will spend or give or lend, yet keepe to have in store,
If fooles may have from hand to mouth, they passe upon no more.

Where ease is sought, at length we see, there plentie waxeth scant,
Who careles lives go borow must, or else full often want.

The world doth think the welthy man is he that least shall need,
But true it is the godlie man is he that best shall speed.

POSIES FOR THE PARLER

AS hatred is the serpents noisome rod,
So friendship is the loving gift of God.

The dronken friend is friendship very evill,
The frantike friend is friendship for the Devill.

The quiet friend all one in word and deede
Great comfort is, like ready gold at neede.

With bralling fooles that wrall for everie wrong,
Firme friendship never can continue long.

In time that man shall seldome friendship mis,
That waith what thing touch kept in friendship is.

Oft times a friend is got with easie cost,
Which used evill is oft as quickly lost.

Hast thou a friend, as hart may wish at will?
Then use him so to have his friendship still.

Wouldst have a friend, wouldst knowe what friend is best?
Have God thy friend, who passeth all the rest.

POSIES FOR THE GESTS CHAMBER

THE sloven and the carles man, the roinish nothing nice,
To lodge in chamber comely deckt, are seldome suffred twice.

With curteine som make scaberd clene, with coverlet their shoo,
All dirt and mire some wallow bed, as spanniels use to doo.

Though bootes and spurs be nere so foule, what passeth some thereon?
What place they foule, what thing they teare, by tumbling therupon.

Foule male some cast on faire boord, be carpet nere so cleene,
what maners careles maister hath, by knave his man is seene.

Some make the chimnie chamber pot to smell like filthie sink,
Yet who so bold, so soone to say, fough, how these houses stink?

They therefore such as make no force what comly thing they spil,
Must have a cabben like themselves, although against their wil.

But gentlemen will gently doe where gentlenes is sheawd,
Observing this, with love abide, or else hence all beshreawd.

POSIES FOR THINE OWNE BED CHAMBER

WHAT wisdom more, what better life, than pleseth God to send?
what worldly goods, what longer use, than pleseth God to lend?

What better fare than well content, agreeing with thy wealth?
what better gest, than trustie friend, in sicknes and in health?

What better bed than conscience good, to passe the night with sleepe?
what better worke than daily care fro sinne thy selfe to keepe?

What better thought, than think on God and daily him to serve?
what better gift than to the poore that ready be to sterve?

What greater praise of God and man, than mercie for to shew?
who merciles shall mercie finde, that mercie shewes to few?

What worse despaire, than loth to die for feare to go to hell?
what greater faith than trust in God, through Christ in heaven to dwell?

A SONET
to the Ladie Paget

SOME pleasures take, and cannot give,
but onely make poore thanks their shift:
Some meaning well, in debt doo live,
and cannot tell how else to shift.

Some knock and faine would ope the doore,
to learne the vaine good turne to praise:
Some shew good face, and be but poore,
yet have a grace, good fame to raise.

Some owe and give, yet still in det,
and so must live, for aught I knowe:
Some wish to pay, and cannot get,
but night and day still more must owe.

Even so must I, for service past,
Still wish you good while life doth last.

Principall points of religion

TO praie to God continually,
To learne to know him rightfully.
To honour God in Trinitie,
The Trinitie in unitie.
The Father in his majestie,
The Sonne in his humanitie,
The holie Ghosts benignitie,
Three persons, one in Deitie.
To serve him alway holily,
To aske him all thing needfully,
To praise him in all companie,
To love him alway hartilie,
To dread him alway christianlie,
To aske him mercie penitently,
To trust him alway faithfully,
To obey him alway willingly,
To abide him alway patiently,
To thanke him alway thankfully,
To live here alway vertuously,
To use thy neighbour honestly,
To looke for death still presently,
To helpe the poore in miserie,
To hope for heavens felicitie,
To have faith hope and charitie,
To count this life but vanitie:
be points of Christianitie.

THE AUTHORS BELIEFE

THIS is my stedfast Creede, my faith, and all my trust,   *God the*
That in the heavens there is a God, most mightie, milde   *Father.*
   and just.
A God above all gods, a King above all kings,
The Lord of lords, chiefe governour of heaven and
   earthly things.

That power hath of life, of death, of heaven and hell,
That all thing made as pleaseth him, so woonderfull to   *Maker of*
   tell: *Heaven.*
That made the hanging Skies, so deckt with divers
   lights,
Of darknes made the cheerfull daies, and all our restfull
   nights.

That clad this earth with herbe, with trees, and sundrie   *The earth.*
   fruites,
With beast, with bird, both wild and tame, of strange
   and sundrie suites :
That intermixt the same with mines like veines of Ore,
Of silver, golde, of precious stones, and treasures many
   more.

That joyned brookes to dales, to hilles fresh water   *The waters.*
   springs,
With rivers sweete along the meedes, to profit many
   things:
That made the hoarie frosts, the flakie snowes so trim,   *Frost and*
The honie deawes, the blustering windes, to serve as   *snowe.*
   pleaseth him.

That made the surging seas, in course to ebbe and flo,   *The seas.*
That skilfull man with sailing ship, mought travell to
   and fro:
And stored so the same, for mans unthankfull sake,
That every nation under heaven mought thereby profit
   take.

The soul of man.	That gave to man a soule, with reason how to live, That doth to him and all things else, his blessing dailie give: That is not seene, yet seeth how man doth runne his race, Whose dailie workes both good and bad, stand knowne before his face.
Thunder and plagues.	That sendeth thundring claps, like terrours out of hell, That man may know a God there is, that in the heavens doth dwel: That sendeth threatning plagues, to keepe our lives in awe, His benefites if we forget, or do contemne his lawe.
Full of mercie.	That dailie hateth sinne, and loveth vertue well, And is the God of Abraham, Isac, and Israell, That doth displeasure take, when we his lawes offend, And yet amids his heavie wrath, his mercie doth extend.
Christ the Sonne.	This is that Lord of hostes, the father of us all, The maker of what ere was made, my God on whom I call: Which for the love of man, sent downe his onelie sonne, Begot of him before the worldes were any whit begonne.
Christes birth. *Christ, God and man.*	This entred Maries wombe, as faith affirmeth sure, Conceived by the holy Ghost, borne of that virgine pure; This was both God and man, of Jewes the hoped king, And lived here, save onely sinne, like man in everie thing.
Christ, our Messias.	This is that virgins childe, that same most holie Preist, The lamb of God, the prophet great, whom scripture calleth Christ, This that Messias was, of whom the Prophet spake, That should tread down the serpents head and our attonement make.

This Judas did betray, to false dissembling Jewes,   *Christes*
Which unto Pilat being Judge, did falsely him accuse: *passion.*
Who (through that wicked Judge) and of those Jewes
   despight,
Condemned and tormented was, with all the force they
   might.

To living wight more evill, what could such wretches do?
More pearcing wounds, more bitter pains, than they did
   put him to?
They crowned him with thorne, that was the king of
   kings,
That sought to save the soule of man, above all worldly
   things.

This was that Pascall lambe whose love for us so stood,   *Christes*
That on the mount of Calverie, for us did shed his blood: *death.*
Where hanging on the Croese, no shame he did forsake,
Till death given him by pearcing speare, an ende of life
   did make.

This Joseph seeing dead, the bodie thence did crave,   *Christes*
And tooke it forthwith from the crosse, and laid it in his   *buriall.*
   grave,
Downe thence he went to hell, in using there his will,   *Christes*
His power I meane, his slained corps in tumb remaining   *descension.*
   still.

From death to life againe, the third day this did rise,   *Christes*
And seene on earth to his elect, times oft in sundrie wise: *resurrection.*
And after into heaven, ascend he did in sight,   *Christes*
And sitteth on the right hand there, of God the father of   *ascension.*
   might.

	Where for us wretches all, his father he doth pray,
	To have respect unto his death, and put our sinnes away:
Christ shall be our judge.	From thence with sounded trump, which noise all flesh shall dread,
	He shall returne with glorie againe, to judge the quicke and dead.
The Judges sentence.	Then shall that voice be heard, Come, come, ye good to mee,
	Hence, hence to hell you workers evill, where paine shall ever bee:
	This is that loving Christ, whom I my Saviour call,
	And onely put my trust in him, and in none else at all.
God the holy Ghost.	In God the holy Ghost, I firmely do believe,
	Which from the father and the sonne a blessed life doth give,
	Which by the Prophets spake, which doth all comfort send,
	Which I do trust shall be my guide, when this my life shall ende.
The Catholike Church.	A holy catholike Church, on earth I graunt there is,
	And those which frame their lives by that, shall never do amis:
	The head whereof is Christ, his word the chiefest post:
	Preserver of this temple great, is God the holy Ghost.
The Communion of Saints.	I do not doubt there is a multitude of Saints,
	More good is don resembling them, than shewing them our plaints:
	Their faith and workes in Christ, that glorie them did give,
	Which glorie we shall likewise have, if likewise we do live.

At God of heaven there is, forgivenesse of our sinnes, *Forgive-*
Through Christes death, through faith in it, and through *nesse of*
   none other ginnes: *sinnes.*
If we repentant here, his mercie dailie crave,
Through stedfast hope and faith in Christ, forgivenes
   we shall have.

I hope and trust upon the rising of the flesh, *Mans resur-*
This corps of mine that first must die, shall rise againe *rection.*
   afresh:
The soule and bodie even then, in one shall joyned bee,
As Christ did rise from death to life, even so through
   Christ shall wee.

As Christ is glorified, and never more shall die, *Life ever-*
As Christ ascended into heaven, through Christ even so *lasting.*
   shall I:
As Christ I count my head, and I a member of his,
So God I trust for Christes sake, shall settle me in blis.

Thus here we learne of God, that there be persons three,
The Father, Sonne, the holy Ghost, one God in trinitee,
In substance all like one, one God, one Lord, one might,
Whose persons yet we do divide, and so we may by right.

As God the Father is the maker of us all,
So God the Sonne redeemer is, to whom for helpe we call,
And God the holy Ghost, the soule of man doth winne,
By mooving hir to waile for grace, ashamed of hir sinne.

This is that God of gods, whom everie soule should love,
Whom all mens hearts should quake for feare his wrath
   on them to move:
That this same mightie God, above all others chiefe,
Shall save my soule from dolefull Hell, is all my whole
   beliefe.

Of the omnipotencie of God,
and debilitie of man.

O GOD thou glorious God, what god is like to thee?
What life, what strength is like to thine, as al the world
    may see?
The heavens, the earth, the seas, and all thy workes
    therein,
Do shew (to whom thou wouldst to know) what thou
    hast ever bin.

But all the thoughts of man, are bent to wretched evill,
Man doth commit idolatrie bewitched of the Devill.
What evill is left undone, where man may have his will,
Man ever was an hypocrite, and so continues still.

What these 4 principal divels do signifie.

What daily watch is made, the soule of man to slea,
By Lucifer, by Belzabub, Mammon, and Asmodea?
In divelish pride, in wrath, in coveting too much,
In fleshly lust the time is spent, the life of man is such.

The joy that man hath here, is as a sparke of fier,
His acts be like the smoldring smoke, himselfe like dirt
    and mier.
His strength even as a reede, his age much like a flower,
His breth or life is but a puffe, uncertaine everie hower.

But for the holy Ghost, and for his giftes of grace,
The death of Christ, thy mercie great, man were in
    wofull case.
O graunt us therefore Lord, to amend that is amisse,
And when from hence we do depart, to rest with thee in
    blisse.

Eleemosyna prodest homini in vita,
in morte, & post mortem.
Out of S. Augustine.

FOR onely love to God, more Christian like to live, *Of almes*
And for a zeale to helpe the poore, thine almes daily give. *deedes.*
Let gift no glorie looke, nor evill possesse thy minde:
And for a truth these profites three, through almes shalt thou finde.

First here the holy Ghost shall daily through his grace,
Provoke thee to repentant life, Gods mercie to embrace.
Of goods and friends (by death) when thou thy leave must take,
Thine almes deedes shall claspe thy soule, and never it forsake.

When God shall after death, call soone for thine account,
thine alms then through faith in Christ, shal al things els surmount.
But yet for any deede, put thou no trust therein,
but put thy trust in God (through Christ) to pardon thee thy sin.

For else as cackling hen with noise bewraies hir nest,
Even so go thou and blaze thy deeds, and lose thou all the rest.

Malus homo
Out of S. Augustine

OF naughtie man, I read, two sundrie things are ment,
The ton is man, the other naught, which ought him to repent.
The man we ought to love, bicause of much therein,
The evill in him we ought to hate, even as filthie sin.
So doth thy daily sinnes the heavenly Lord offend,
But when thou dost repent the same, his wrath is at an end.

OF TWO SORTS OF MEN,
the tone good, and tother bad,
Out of S. Augustine

SINCE first the world began, there was and shall be still,
Of humane kind two sundrie sorts, thon good and thother ill:
Which till the judgement day, shall here togither dwell,
But then the good shall up to heaven, the bad shall downe to hell.

Diabolo cum resistitur, est ut formica:
Cum vero eius suggestio recipitur, fortis est ut leo.
Out of S. Augustine.

WHEN Sathan we resist, a Pismier shall he be,
But when we seeme to give him place, a Lion then is he.

EIGHT OF S. BARNARDS VERSES
both in Latine and English
with one note to them both

Cur mundus militat, sub vana gloria,
Cuius prosperitas, est transitoria?
Tam cito labitur, eius potentia,
Quam vasa figuli, quæ sunt fragilia?

Why so triumphes the world, in pompe and glorie vane,
Whose state so happie thought, so fickle doth remaine?
Whose braverie slipprie stands, and doth so soone decaie,
As doth the potters pan, compact of brittle claie?

Plus crede literis, scriptis in glacie,
Quam mundi fragilis, vanæ fallaciæ,
Fallax in præmiis, virtutis specie,
Quæ nunquam habuit tempus fiduciæ.

More credite see thou give, to letters wrote in ise,
Than unto vaine deceits, of brittle worlds devise.
In gifts to vertue due, beguiling many one,
Yet those same never have long time to hope upon.

Magis credendum est, viris fallacibus,
Quam mundi miseris prosperitatibus,
Falsis insaniis et voluptatibus,
Falsis quoque studiis et vanitatibus.

To false dissembling men more trust is to be had,
Than to the prosperous state of wretched world so bad:
What with voluptuousnes, and other maddish toies,
False studies won with paine, false vanities and joies.

Dic ubi Salomon, olim tam nobilis?
Vel ubi Samson est, dux invincibilis?
Vel dulcis Ionathas, multum amabilis?
Vel pulcher Absolon, vultu mirabilis?

Tell where is *Salomon*, that once so noble was?
Or where now *Samson* is, in strength whome none could pas?

Or woorthie *Ionathas*, that prince so lovely bold?
Or faier *Absolon*, so goodlie to behold?

> *Quo Cæsar abiit, celsus imperio?*
> *Vel Dives splendidus, totus in prandio?*
> *Dic ubi Tullius, clarus eloquio?*
> *Vel Aristoteles, summus ingenio?*

Shew whither in *Cæsar* gone, which conquered far and neere?
Or that rich famous *Carle*, so given to bellie cheere:
Shew where is *Tullie* now, for eloquence so fit?
Or *Aristoteles*, of such a pregnant wit?

> *O esca vermium! o massa pulveris!*
> *O ros! o vanitas! cur sic extolleris,*
> *Ignoras penitus utrum cras vixeris,*
> *Fac bonum omnibus, quam diu poteris.*

O thou fit bait for wormes! O thou great heape of dust!
O dewe! O vanitie! why so extolst thy lust?
Thou therefore ignorant, what time thou hast to live,
Doe good to erie man while here thou hast to give.

> *Quam breve festum est, hæc mundi gloria?*
> *Ut umbra hominis, sic eius gaudia,*
> *Quæ semper subtrahit, æterna præmia,*
> *Et ducunt hominem, ad dura devia.*

How short a feast (to count) is this same worlds renowne?
Such as mens shadowes be, such joies it brings to towne.
Which alway plucketh us from Gods eternall blis:
And leadeth man to hell, a just reward of his.

> *Hæc mundi gloria, quæ magni penditur,*
> *Sacris in literis, flos fœni dicitur,*
> *Ut leve folium, quod vento rapitur,*
> *Sic vita hominum, hac vita tollitur.*

The braverie of this world, esteemed here so much,
In Scripture likened is, to flowre of grasse and such:
Like as the leafe so light, through winde abrode is blowne,
So life in this our life, full soone is overthrowne.

OF THE AUTHORS LINKED VERSES
departing from Court to the Countrie

Muse not my friend to finde me here,
Contented with this meane estate:
And seeme to doo with willing cheere,
That courtier doth so deadly hate.

For fortunes looke,
Hath changed hew:
And I my booke,
Must learne anew.

And yet of force, to learne anew,
Would much abash the dulled braine:
I crave to judge if this be trew,
The truant child that knowth the paine.

But where a spight,
Of force must bee:
What is that wight,
May disagree?

No, no, God wot, to disagree,
Is ventring all to make or mar:
If fortune frowne we dailie see,
It is not best to strive too far.

For lordlie bent,
Must learne to spare:
And be content
With countrie fare.

From daintie Court to countrie fare,
Too daintie fed is diet strange:
From cities joy, to countrie care,
To skillesse folke is homelie change.

Where neede yet can,
None other skill:
Somtime poore man
Must breake his will.

If courtlie change so breaketh will
That countrie life must serve the turne:
What profit then in striving still,
Against the prick to seeme to spurne?

If court with cart
Must be content,
What ease to hart,
Though mind repent?

What gaine I though I doo repent,
My crotches all are broke and gon:
My woonted friends are careles bent,
They feare no chance I chance upon.

As neede doth make
Old age to trot:
So must I take,
In woorth my lot.

Now if I take in woorth my lot,
That fatall chance doth force me to,
If ye be friends embraid me not,
But use a friend as friends should do.

Behold the horse
Must trudge for pelfe,
And yet of forse,
Content it selfe.

THE AUTHORS LIFE

Epodium.   NOW gentle friend, if thou be kinde,
            Disdaine thou not, although the lot
            Will now with me no better be,
               than doth appere:
            Nor let it grieve, that thus I live,
            But rather gesse, for quietnesse,
            As others do, so do I to,
               content me here.

            By leave and love, of God above,
            I minde to shew, in verses few,
            How through the breers, my youthfull yeeres,
               have runne their race:
            And further say, why thus I stay,
            And minde to live, as Bee in hive,
            Full bent to spend my life to an end,
               in this same place.

Borne at   It came to pas, that borne I was
Rivenhal   Of linage good, of gentle blood
in Essex.   In Essex laier, in village faier,
               that Rivenhall hight:
            Which village lide by Banketree side,
            There spend did I mine infancie,
            There then my name, in honest fame,
               remaind in sight.

            I yet but yong, no speech of tong,
            Nor teares withall, that often fall
Set to song   From mothers eies, when childe out cries,
schoole.   to part hir fro:
            Could pitie make, good father take,
            But out I must, to song be thrust,
            Say what I would, do what I could,
               his minde was so.

O painfull time, for everie crime,	Queristers
What toesed eares, like baited beares?	miserie.
What bobbed lips? what jerks, what nips?	
what hellish toies?	
What robes, how bare? what colledge fare?	
What bread, how stale? what pennie ale?	
Then Wallingford, how wart thou abhord	Wallingford
of sillie boies?	Colledge.

Thence for my voice, I must (no choice)	
Away of forse, like posting horse,	
For sundrie men, had plagards then,	Singing
such childe to take:	mens com-
The better brest, the lesser rest,	missions.
To serve the Queere, now there now heere,	
For time so spent, I may repent,	
and sorrow make.	

But marke the chance, my self to vance,	
By friendships lot, to Paules I got,	
So found I grace, a certaine space,	John Redford
still to remaine:	an excellent
With Redford there, the like no where,	Musician
For cunning such, and vertue much,	[organist of
By whom some part of Musicke art,	St. Paul's. M.]
so did I gaine.	

From Paules I went, to Eaton sent,	Nicholas
To learn streight waies, the latin phraies,	Udall
Where fiftie three stripes given to mee,	school-
at once I had:	master at
For fault but small, or none at all,	Eaton.
It came to pas, thus beat I was,	
See Udall see, the mercie of thee,	
to me poore lad.	

Trinitie hall in Cam-bridge.

From London hence, to Cambridge thence,
With thankes to thee, O Trinitee,
That to thy hall, so passing all,
   I got at last:
There joy I felt, there trim I dwelt,
There heaven from hell, I shifted well,
With learned men, a number then,
   the time I past.

Quartan ague.

Long sicknes had, then was I glad
To leave my booke, to prove and looke,
In Court what gaine, by taking paine,
   mought well be found:

Lord Paget good to his servants.

Lord Paget than, that noble man,
Whose soule I trust is with the just,
That same was hee enriched mee,
   with many a pound.

When this betide, good parents dide,
One after one, till both were gone,
Whose petigree, who list may see,
   in Harolds Booke:
Whose soules in blis be long ere this,

The hope we have of the dead.

For hope we must, as God is just,
So here that crave shall mercie have,
   that mercie looke.

The vices of the Court.

By Court I spide, and ten yeres tride
That Cards and Dice, with Venus vice,
And peevish pride, from vertue wide,
   with some so wraught:
That Tiburne play made them away,
Or beggers state as evill to hate,
But such like evils, I saw such drevils,
   to come to naught.

Yet is it not to be forgot,
In Court that some to worship come,
And some in time to honour clime,
   and speede full well:
Some have such gift, that trim they shift,   *The court*
Some profite make, by paines they take,   *commended.*
In perill much, though oft are such,
   in Court that dwell.

When court gan frowne and strife in towne,   *The nobilitie*
And lords and knights, saw heavie sights,   *at variance*
Then tooke I wife, and led my life   *in Edward*
   in Suffolke soile.   *the 6 daies.*
There was I faine my selfe to traine,   *Katewade.*
To learne too long the fermers song,
For hope of pelfe, like worldly elfe,
   to moile and toile.

As in this booke, who list to looke,   *At Kate-*
Of husbandrie, and huswiferie,   *wade in*
There may he finde more of my minde,   *Suffolke this*
   concerning this: *booke first*
To carke and care, and ever bare,   *devised.*
With losse and paine, to little gaine,
All this to have, to cram sir knave,
   what life it is.

When wife could not, through sicknes got,
More toile abide, so nigh Sea side,
Then thought I best, from toile to rest,   *Ipswich*
   and Ipswich trie: *commended.*
A towne of price, like paradice,
For quiet then, and honest men,
There was I glad, much friendship had,
   a time to lie.

The deth of his first wife.	There left good wife this present life, And there left I, house charges lie, For glad was he, mought send for me, good lucke so stood: In Suffolke there, were everie where, Even of the best, besides the rest, That never did their friendship hid, to doo me good.
Newe maried in Norfolk.	O Suffolke thow, content thee now, That hadst the praies in those same daies, For Squiers and Knights, that well delights good house to keepe: For Norfolke wiles, so full of giles, Have caught my toe, by wiving so, That out to thee, I see for mee, no waie to creepe.
Mistres Amie Moone.	For lo, through gile, what haps the while, Through Venus toies, in hope of joies, I chanced soone to finde a Moone, of cheerfull hew: Which well a fine me thought did shine, Did never change, a thing most strange, Yet kept in sight, hir course aright, and compas trew.
The charges following a yoong wife.	Behold of truth, with wife in youth, For joie at large, what daily charge, Through childrens hap, what opened gap, to more begun. The childe at nurse, to rob the purse. The same to wed, to trouble hed. For pleasure rare, such endlesse care, hath husband wun.

Then did I dwell in Diram sell,   *West*
A place for wood, that trimlie stood,   *Diram*
With flesh and fish, as heart would wish: *Abbie.*
   but when I spide
That Lord with Lord could not accord,   *Land-lordes*
But now pound he, and now pound we,   *at variance.*
Then left I all, bicause such brall,
   I list not bide.

O Soothwell, what meanst thou by that,   *Sir Richarde*
Thou worthie wight, thou famous knight,   *Soothwell.*
So me to crave, and to thy grave,
   go by and by?
O death thou fo, why didst thou so
Ungently treat that Jewell great,
Which opte his doore to rich and poore,
   so bounteously?

There thus bestad, when leave I had,
By death of him, to sinke or swim,
And ravens I saw togither draw,   *His vii*
   in such a sort: *executors.*
Then waies I saught, by wisdome taught,
To beare low saile, least stock should quaile,
Till ship mought finde, with prosperous winde,
   some safer port.

At length by vew, to shore I drew,
Discharging straight both ship and fraight,
At Norwich fine, for me and mine,   *Norwich*
   a citie trim: *Citie.*
Where strangers wel may seeme to dwel,   *Norwich*
That pitch and pay, or keepe their day,   *qualities.*
But who that want, shall find it scant
   so good for him.

*Maister   * But Salisburie how were kept my vow,
Salisburie If praise from thee were kept by mee,
deane of Thou gentle deane, mine onely meane,
Norwich.   there then to live?
          Though churles such some to crave can come,
          And pray once got, regard thee not,
          Yet live or die, so will not I,
             example give.

In 138 When learned men could there nor then,
houres I Devise to swage the stormie rage,
never made Nor yet the furie of my dissurie,
drop of   that long I had:
water. From Norwich aire, in great despaire,
          Away to flie, or else to die,
          To seeke more helth, to seeke more welth,
             then was I glad.

          From thence so sent, away I went,
          With sicknes worne, as one forlorne,
Fairsted To house my hed, at Faiersted,
parsonage   where whiles I dwelt:
in Essex. The tithing life, the tithing strife,
          Through tithing ill, of Jacke and Gill,
          The dailie paies, the mierie waies,
             too long I felt.

          When charges grew, still new and new,
          And that I spide, if parson dide,
Lease for (All hope in vaine) to hope for gaine,
parsons life.   I might go daunce:
          Once rid my hand of parsonage land,
          Thence by and by, away went I,
          To London streight, to hope and waight,
             for better chaunce.

Well London well, that bearst the bell   *London*
Of praise about, England throughout,   *commended.*
And doost in deede, to such as neede,
    much kindnes shew:
Who that with thee can hardly agree,
Nor can well prais thy friendly wais,
Shall friendship find, to please his mind,
    in places few.

As for such mates, as vertue hates,   *Unthrifts*
Or he or thay, that go so gay,   *order.*
That needes he must take all of trust,
    for him and his:
Though such for wo by Lothburie go,
For being spide about Cheapeside,
Least Mercers bookes for monie lookes,
    small matter it is.

When gaines was gon, and yeres grew on,   *The plague*
And death did crie, from London flie,   *at London*
In Cambridge then, I found agen,   *[1574,1575]*
    a resting plot:
In Colledge best of all the rest,   *Trinitie*
With thankes to thee, O Trinitee,   *College in*
Through thee and thine, for me and mine,   *Cambridge.*
    some stay I got.

Since hap haps so, let toiling go,
Let serving paines yeeld forth hir gaines,
Let courtly giftes, with wedding shiftes,   *Youth ill*
    helpe now to live: *spent makes*
Let Musicke win, let stocke come in,   *age repent.*
Let wisedome kerve, let reason serve,
For here I crave such end to have,
    as God shall give.

A lesson for Thus friends, by me perceive may ye,
yonger   That gentrie standes, not all by landes,
brothers. Nor all so feft, or plentie left
      by parents gift:
     But now and then, of gentlemen,
     The yonger sonne is driven to ronne,
     And glad to seeke from creeke to creeke,
      to come by thrift.

A true   And more by this, to conster is,
lesson.   In world is set, ynough to get,
     But where and whan, that scarsely can,
      the wisest tell:
     By learning some to riches come,
     By ship and plough some get ynough,
     And some so wive that trim they thrive,
      and speede full well.

Hardnes in To this before, adde one thing more,
youth not Youth hardnes taught, with knowledge wraught,
the worst. Most apt do proove, to shift and shoove,
      among the best:
Cocking of Where cocking Dáds make sawsie lads,
youth not In youth so rage, to beg in age,
the best. Or else to fetch a Tiburne stretch,
      among the rest.

     Not rampish toie, of girle and boie,
     Nor garment trim, of hir or him,
     In childhoode spent, to fond intent,
      good end doth frame:
Not pride in If marke we shall, the summe of all,
youth, but The end it is, that noted is,
welth in age Which if it bide, with vertue tride,
needfull.   deserveth fame.

When all is done, lerne this my sonne,
Not friend, nor skill, nor wit at will,
Nor ship nor clod, but onelie God,
   doth all in all:
Man taketh paine, God giveth gaine,
Man doth his best, God doth the rest,
Man well intendes, God foizon sendes,
   else want he shall.

Man doth labour and God doth blesse.

Some seeke for welth, I seeke my helth,
Some seeke to please, I seeke mine ease,
Some seeke to save, I seeke to have
   to live upright:
More than to ride, with pompe and pride,
Or for to jet, in others det,
Such is my skill, and shall be still,
   for any wight.

A contented minde is worth all.

Too fond were I, here thus to lie,
Unles that welth mought further helth,
And profit some should thereby come,
   to helpe withall:
This causeth mee well pleasde to bee,
Such drift to make, such life to take,
Enforsing minde remorse to finde,
   as neede neede shall.

Friend, al thing waid, that here is said,
And being got, that paies the shot,
Me thinke of right have leave I might,
   (death drawing neere:)
To seeke some waies, my God to praies,
And mercie crave, in time to have,
And for the rest, what he thinkes best,
   to suffer heere.

Happie that lives well, unhappie that dies evill.

>Fortuna non est semper amica,
>Superbian igitur semper devita.

(*This poem is only in the editions of 1573 and 1577.*)

Though Fortune smiles, and fawnes upon thy side,
　　Thyself extol for that no whit the more:
Though Fortune frownes and wresteth al thing wide,
　　Let fancy stay, keepe courage still in store;
　　For chance may change as chance hath don before:
Thus shalt thou holde more safe then honour got,
Or lose the losse, though Fortune will or not.

Thy friend at this shall dayly comfort have,
　　When warely thus, thou bearest thy selfe upright,
Thy foes at this shall gladly friendship crave,
　　When hope so small is left to wrecke their spight,
　　For lowly liefe withstandeth envy quight:
As floeting ship, by bearing sayl alowe,
Withstandeth stormes when boistrous winds do blowe.

Thy usage thus in time shall win the gole,
　　Though doughtful haps, dame fortune sendes betweene,
And thou shalt see thine enemies blow the cole,
　　To ease thine hart much more then thou dost weene,
　　Ye though a change most strangely should be seene,
Yet friend at neede shall secret friendship make,
When foe in deede shal want his part to take.

FINIS

NOTES AND ILLUSTRATIONS

Abbreviations used: H—Hillman, *Tusser Redivivus* 1710; M—Tusser, *Five Hundred Points* ed. Mavor 1812; E.D.S. Tusser *Five Hundred Points* ed. English Dialect Society 1878.

Page 1.2. "Remaine abrode for evermore"—be given to the writings of others. E.D.S.

Page 4.3. "That, that waie nothing geanie"—he kept so many servants that he had no gains left for himself. M.

Page 5.1. "Ictus sapit"—corresponds to our proverb, "The burnt child dreads the fire". E.D.S.

Page 7.4. If Tusser speaks literally, the price of his book was only a groat, or two at the utmost. M.

Page 13. "Familie"—household, servants. E.D.S.

Page 16.4. "Of tone of them both"—the one or them both. M.

Page 17.7. "Be to count ye wote what"—of little importance. E.D.S.

Page 18.6. "Skarborow warning"—A sudden surprise when a mischief is felt before it is suspected. This proverb takes its origin from Thomas Stafford, who, in 1557, with a small company, seized on Scarborough Castle before the Townsmen had the least notice of his approach. Ray.

Page 18.7. "Legem pone"—ready money. "Oremus"—to beg, here making excuses for non-payment of debts. "Praesta quaesumus"—lend me, I pray. E.D.S.

Page 18.8. "Nor put to thy hand," *etc.*—do not meddle in family affairs. E.D.S.

Page 20.2. "Docking the dell"—dissipation. Grose's *Dictionary*.

Page 24.2. "Veale and Bakon is the man"—is the proper food or is in season.

"Martilmas beef" is killed at Martinmas [11 November] and dried in the chimney like bacon. E.D.S.

Page 24.3. The Lamb and their wool commonly pays their price and their going, and the Country Man has a Carcass of very good Mutton for nothing, and sometimes less than nothing: but his Care and Skill is required in buying in at the first. H.

Page 24.4. "All Saints doe laie, *etc.*"—All Saints' Day expects or lays itself out for pork and souse (pickled pigs' feet and ears) sprats and smelts for the household. E.D.S.

Page 24.6. So that here is to be noted, altho' our Author was a very sound Protestant as appears by his Belief, and several other writings of his: yet he thought it no Popery to keep the Ember Weeks, the Vigils, (which I take to be what he means by Fasting-Days) and *Fridays, Saturdays* and *Wednesdays* as Days of Fasting and Abstinence: and not only he, but that it was the Custom of the Times wherein he liv'd, the Custom in Queen *Elizabeth's* Days, in which this his Book was publish'd. Neither is there any reason that a good Custom should be utterly abolish'd because it has been abused, or because Men err in some things, they must be supposed to do so in everything. But it seems the desire of Novelty had not yet so much intoxicated Men as it has done since our Author, and with him the Farmer-like part of the Nation had their set Feasting Days also, as follows: which if he had thought was superstitious, he would hardly have recommended. H.

Page 30.1. So that this Month or the Observations on it, are suppos'd not to begin until Michaelmas Even, that is, when a Farmer first comes into a new Farm. But seeing in some Farms they are obliged to Summer-fallow their Grounds with their Neighbours, it is unreasonable that the new Farmer should lose the Benefit of that Ground for that Year, which he must do, unless he can Summer-fallow when others do. H.

Page 30.2. But this and all other Conveniences are best provided for by Lease, for it is but a sorry Plea to plead Custom against one that is in Possession, and can make what Customs he pleases. H.

The Off-going tennant of champion or open field allows the incoming to summer fallow that portion of the ground destined for wheat.

But the occupier of woodland or enclosures holds the whole till the expiration of his term, unless particular stipulations are made by lease: and without a lease neither the real interest of the tenant nor the landowner can be consulted. M.

Page 30.3. A sure Bargain hinders all Contention, and as no body can blame a Farmer for using all his Wit and Cunning in taking a Farm, so neither ought a Gentleman be blam'd for using all his in the letting it: and it is very rare that either of them get any thing by ambiguous Terms, which serve for nothing but to nourish Strife, and in the end do fraud both. Here the Author reckons up the Twelve Properties of a good Farm. H.

Page 30.4. A good Farm is such a one as bears a due Proportion between the whole and its parts, as if it be a Corn Farm, that it have a due proportion of Meadow and Pasture, that its Sheep-walk be not under stinted, that Commonage lie convenient, that Dung, Chalk, or Marle may be had, that there be no Scarcity of Firing, Plow-Boot and Cart-Boot (Wood to mend Plow and Cart,) and that the Rent be not over dear. H.

Page 30.5. This is no more than to take care, first of your Grounds, then of your Dwelling, to shew that the one is more material than the other, and that the latter may have another time, but a delay in the former is more difficult, if not impossible to retrieve: however, they are both needful. H.

Page 30.6. If you do, assure your self, your Stock will be sinking: and Old certainly grows Older, and for the most part little degenerates to less. If they should thrive with you, it is a question whether they will pay their Pasture, considering how long you must keep them, and how much you are behind the Market. H.

Page 31.1. The lightness here spoken of, is a cleaverliness, a proportionate Strength, for a Horse or an Ox is neither so healthy or fit for Service when he is loaden with a Mass of Flesh, as when he is between both, in what the Farmers call good Tune. Neither is it a great thickness of Timber that makes any thing strong, especially such things as are to be in motion, as a Plough, Cart, Waggon, etc., but a due and near compactness wherein every thing is made fit for the Work it is design'd for, and not burthen'd with its own Weight. But the Soil, Seed, Sheaf and Purse our Author excepts, altho' the Soil and Purse may be too heavy if they contain matter of little value. H.

Page 31.3. He begins with the Barn Utensils, of which many are so well known they need no description, however, the repeating them is a good Remembrancer, as for example, Barn lock'd, I take to mean that the Farmer should see carefully to the locking up his Barns, which if he does not, he shall find an out-let, that he may be sensible of before he has paid his Rent. What of his own Horses are the receivers, the Theft is not the less, but sometimes the more: for what they get that way, the Servant will have out of them again in hurrying. A Gofe is what in some places is called a Mow, to which there belong a Ladder for the Thresher to get up and throw down the Corn, a short pitchfork for that use, and a long one to pitch up the Straw when his Straw-Mow grows: high a Strawfork and Rake to turn the Straw off from the thresh'd Corn, a Fan and a Wing to clean it (which by the way is much better than meer winnowing with a Wind Winch) as giving the Corn a brighter Colour, and freeing it from Dust. A Cart Nave I suppose is to stand up upon when they Wind-winnow, a casting Shovel is such as Malt-Men use, and serves to cast Wheat or Beans the length of the Floor, and thus Seed-Wheat should be serv'd, for the best Grains fly farthest, and may be thus separated from the lighter. H.

Page 31.4. Planking of Stables is by some not so well approv'd on as pitching. However, the meaning here is, that the Horse lie dry and sweet, as to his making sweet Chaff a Stable utensil. It is very useful and proper to mix Chaff with the Oats the Farmer gives his labouring Horses, it not only fills and affords a good and dry nourishment, but the Horse eats the Oats mixt with them the better, for finding them of better taste than the Chaff, he strives to chew them, which for greediness when he has clear Oats he does not, but swallows many whole. H.

Page 31.5. A Skep is a sort of Basket (with a wooden handle. M.), narrow at the bottom, and wide at the top, to fetch Corn in. A Bin is a small enclos'd place in some corner, to put Oats, Chaff or Beans. H.

Page 31.6 A Buttrice is what the Farriers pare Horses Hoofs withal, which he would have his Farmer provided with, as well as with Pincers, Hammer, Nails and Apron, that he may not be forced to go to a Farrier for every small matter. A Nall is an Awl such as Collar makers use, which he would also have his Farmer provided with, as well as some other Tools of that Trade, particularly Whit-Leather to mend his Collars and Harness when there is any occasion. H.

Page 31.7. A Pannel and Ped have this difference, the one is much shorter than the other, and rais'd before and behind, and serves for small Burdens, as the Maid to Market with her Butter, the Boy to Mill. The other is longer, and made for Burdens of Corn, and is most in use, where Wains and Carts cannot travel: these are fastened with a leathern girt, call'd a Wantye. A Packsaddle is not so frequently found among Farmers as formerly, except in the Northren Parts, where it is used to carry Wooll. A Chain for a Stable is of good use, both to balk a Thief, who, when he has broke open your Door, will be ne're the nearer. And to keep your Horses in whilst you are harnessing them, and receive light from the Door. H.

Page 32.1. Clouting a Wheel is arming the Axle-tree with Iron Plates, to keep it from wearing. Shod is arming the Fellows with Iron Strakes, or a Tire as some call it on the outward Circumference of the Wheel. Cart Ladders and Wheel Ladders are Frames on the sides and Tail, to support light Loads, as Hay, etc. A Pod I take to be a Box, or some old leather Bottle nail'd to the side of the Cart, to hold the Percer, Wimble and Nails and Hammer if need be, altho' that is often adraught Pin for the Thiller or hindmost Horse. Shaving a Whiplath, is shaving of a tough piece of Whitleather thin, for the Lath of a Cart-Whip. H.

Page 32.2. A Coome is four Bushels, and forty Bushels is a Load of Wheat. His pulling Hook is a barbed Iron, by the help of which, short Bushes, Broom, Brakes and other light firing may be pull'd out of the Stack without hurting the Hands. A Tumbrel is a dung Cart, and sometimes used for other uses. A dung Crone is a dung Hook, wherewith Dung is unloaden. H.

Page 32.4. A Nads is an Adz, used by Carpenters to even flooring, and may also serve very well to hollow a Hog-Trough. H.

Here are some more odd things, amongst which the most remarkable is a Frower of Iron, for cleaving of Lath. Now this Lath must be for the Farmer's own use, for it is not to be suppos'd that the Landlords of those days allow'd the Tenants to fell their Timber, whether converted into Lath or otherwise: so that I take it to be for the sides of their Carts, Waggons or Waines, which still in some places is in use, and perhaps for airings of their Barns, etc. N.B. Because it is here called a Lath, it follows not that it was no thicker than our ordinary Lath is at present, for all that is split may go under that denomination, and perhaps Pales are hereby meant also. H.

"Dovercourt beetle"—a mallet made of the wood of the celebrated elms of Dovercourt in Essex. Harrison (Descr. of Engl.) states that these trees supplied the most durable wood to be found in England. [Hilman states that a "Dovercourt beetle" is a "very large beetle," which makes a great noise. This is, of course, incorrect.]

Page 32.5. As to two Ploughs, they may be necessary, because it is very likely the same Farm may require two sorts: namely, a Wheel-Plough for Stony, and a Swing Plough for Clay: but why three Shares I know not. Ground and side Clouts may be made of old Streaks of Wheels, which ought carefully to be saved for this and other purposes, as indeed everything ought so to be, that it is probable will be of any use. H.

Page 32.6. Strutt (Manners and Customs ii. 12) alludes to the ploughstaff: 'The ploughman holdeth the ploughstilt [i.e. principal handle] in his left hand, and in his right the ploughstaff to break the clods.' Breaking the Clods after the Plough, here we find of ancient use: it is pity it is not continued, for that will break them when new turn'd up, which must sometimes lie a long time to mellow with the Weather. H.

Page 32.7. Sedge Collars are by much the lightest and coolest, indeed not so comly as those of Wadmus, but will serve a good Team well enough to go to plough with.

Well clad is here brought in for rhime sake, and signifies in good tune or good heart: not that plough Horses should be kept in Horse Cloths. H.

Page 32.8. This sort of barly Rake is still us'd in Norfolk and Suffolk, and is drag'd by a lusty Fellow all over the Ground after the Cocks are taken off, and gathers a great deal better than a hand Rake, if the Ground has been well roll'd.

A Mother or Mather is a young Wench, for whom our Author thinks a Sling more proper than a Bow, which he assigns the Boy. These were made use of for driving away Crows from the Corn, and which perhaps is the reason why Bows came to be so frequent at Bartholemew Fair. H. "Hoigh de la roy"—excellent or proper. E.D.S.

Page 33.1. A brush Sithe I take to be an old Sithe to cut up Weeds, as Nettles, Hemlock, etc. Some use a wooden Sithe to kill Fern, and the Weed call'd Kedlack or Carlake, when they grow among Corn. The brushing of their tops hurts not the Corn at all, and they dye a good way after the Sithe, so that the Corn soon overshoots them. A Rifle or Rufle is no more than a bent Stick standing on the butt of Sithe handle, by which the Corn is struck together in Rows. A Cradle is a three forked Instrument of Wood, on which the Corn is caught as it falls from the Sithe, and laid more regularly than otherwise it would be. This lies very heavy on the hand, and therefore much disus'd: however, for ought I know, it might save abundance of Labour in our Northren Parts, where they reap their Barley, Oats and Bigg. A Meath is a Hook at the end of a handle five Foot long, with which, formerly Pease were cut, by now left off, and a short Sithe us'd for the most part. H.

Page 33.2. The short Rake is well known, of the long Rake has been mentioned enough already, which for Barley may have wooden Teeth, but the Rake to rake up the Fitches that lie, is the Iron-tooth'd Rake before mention'd, which tears away what has been left uncut behind. A pike is no other than a Pitchfork, with three Tines, such as Barley, Oats, etc. are generally cock'd with. H.

Page 33.3. A Scuttle is the same as a Skep, altho' this may be suppos'd a larger one than that of the Stable. The Fork and Hook to be tempering Clay, are a three tin'd Fork, the same with a Dung Fork, and the Hook what he call'd before a Crone, their use is to mix straw Loam, or Clay for Loam Walls, for which also is the Lath, Hammer, Trowel and Hod. H.

Page 33.4. Twitchers are a sort of great Plyers to clinch the Hog-Ring withal. Tar is the Husband Mans Oyntment which he applies outwardly to all wounds of Sheep and Hogs, and sometimes gives it inwardly. Two pints of Tar to a pound of Pitch is the Composition still kept up for Sheep-Marks. H.

Page 33.6. A Skuppat is a sort of Scoop or hollow Shovel in use with Mersh Men, to throw out Water or thin Mud out of Ditches. A Skavel is a sort of Spade about four Inches wide at the bottom, and eight Inches deep, to cut Earth out of the solid when new Ditches are made, and where the throw is any thing considerable. The Sickle here spoken of is a Hook at the end of a 10 or 12 foot Pole, to cut Weeds at the bottom of a Drain. A Didal is a triangular Spade, as sharp as a Knife: excellent to bunk Ditches where the Earth is light and pester'd with a sedgy Weed, Workmen call a Dag-prick. A Crome is like a Dung-Rake, with a very long Handle, to pull Weeds out of a Drain after they are cut. H.

Page 33.7. Hilman supposes that the "Soles" here mentioned refer to "Soles for Shoes, which he would have every Farmer to have in readiness, to sell to his Servants when they want them." This is wrong, however, for "Sole" is a collar of wood put on cattle to confine them to the post. E.D.S., *Glos*.

Page 34.1. Last Years Corn will grow, but is longer coming up, more apt to burst: and (because more die in the Ground) requires more Seed than new, so that without doubt new Seed is the best: also because if thresh'd before it has throughly sweated in the Mow, the thin Corn will stay in the Ear, and none but the best Corn will come out. Hence some slash their Wheat and Rye Sheaves upon a Hurdle for Seed-striking is the last plowing before sowing, when sowed above furrow, and if the Ground be cloddy, to be sure it is good to break them. H.

Page 34.2. Let them be out of the Milk before the Frost come if possible, and have a full threaded Root, and they will take little harm: unless the ensuing Frosts are very black and hard indeed. H.

Page 34.3. Tems Bread is that out of which the courser Bran is taken, and is somewhat finer than ordinary Farmers use. This may be very good, altho' some Rye be mix'd with it, nay, to most Palates it is more grateful than Wheat alone: because it retains a Moisture, so that Wheat and Rye mix very well in Bread. But our Author is by no means to have them mixed in Seed, altho' some sort does pretty well. H.

Page 34.4. Because white Wheat will grow on a lighter Mould than red Wheat. H.

Page 34.5. Beans are a strong Pulse, and have a broad Leaf, with which they drip the Weeds more than either Wheat, Rye or Pease: however, in this the Nature of the Ground, and what it is able to bear, is to be consider'd: and notwithstanding their strength, Beans thrive best when weeded, either with the Hoe or Hand, as doth all other Corn. Dredge is a mixture of Oats and Barly, now very little sown. H.

Page 34.6. And Reason good, for these Lawless Thieves are cherish'd in such numbers, that they are one of the Farmers greatest Plagues: I have heard, modestly computed, that a pair of Pidgeons will starve on a quarter of Corn in a Year, and the Rook watches the first sprouting of the Corn more nicely than the Farmer can. H.

Page 34.7. A Water-Furrow runs cross the Ridges most commonly, and is always made in the lowest part of the Land. The Dyking up ends of Common-Field-Land against the Highways, will do somewhat where there is no other means to fence your Ground, but it is a very weak defence. H.

Page 34.8. A Mersh-Wall is a Sea-bank, made with a considerable slope to Sea-ward, which is called a Break, or Breck: it is faced with Turf, which sometimes is worn by the Sea, or Holes made in it by Crabs, etc. The Foreland is a piece of Land that lies from the foot of the Bank to seaward, and must be well look'd after, that it wear not away, or come too near the Bank (as the Workmen term it) and this before Michaelmas, for the Tydes near the Autumnal Equinox are most outragious. H.

Page 35.1. The Nights are now moderately cold, and Beasts in pretty good heart, and Leaping-time over, makes it the best time to geld in. All Fruit intended to be kept, must be gathered dry: and Walnuts no less than any other, for if their outward Husk rot, the Nutshel will be black. H.

Page 35.2. As to any Astrological Observation why Fruit should last that is gathered in the Wane of the Moon, I leave it to the more Learned: But this I know, that at this time of the Year, after the Wane, the fore-parts of the Nights are dark, and the Mornings Moonshine: of this perhaps the Michers, as our Author calls Thieves may take some advantage: and certainly the way to gather Fruit to last, is to get it in before it is gone. H.

Page 35.3. He might have added if Fruit stand too long it will be mealy, which is worse than shrively, for now most Gentlemen chuse the shriveled Apple. H.

Page 35.4. What are taken at this time of the Year, must be killed: the best way is to suffocate them with Brimstone: and what are drove at Mid-Summer, seldom live over the Winter: so that the Cruelty used towards them is much alike. There have been many ingenious Ways contriv'd to save the Life of this Creature, and I should be glad to hear of any that turn to account, what do not are the same as if the Farmer should keep his Ox and his Sheep beyond their Prime, and lose the profit of their Flesh, for the Labour of the one and the Wooll of the other. H.

Page 35.5. That is, it must stand above the Grass and Weeds, for the natural Defect of this Creature is short-sightedness, and when they come Home loaden, whatever is above the Stool incumbreth them, and if they pitch amongst thick Grass, they are not able to rise again. H.

Page 35.6. There is now very little Feed for him in the Fields, and if he get into the Woods, he will follow the first Sow he finds with Brim: and being entertain'd every where if he be but out of knowledge, you may have him a good way to seek. Hard and cool lying makes him rub stoutly, which increases his Shield: (as the Skin of the Shoulder is called). H.

Page 35.6. Measles in a Hog are little round Globules that lie amongst the Muscles; they are known to be occasion'd thro' want of Water, perhaps the Chyle thereby is too thick, and unapt to be turn'd into pure Blood. H.

Page 35.7. The retting of Hemp is commonly done in standing Plashes, or small Pools, on Commons near Roads, etc. and must be watched, and taken out as soon as it begins to swim. It leaves a loathsome Smell in the Water. H.

Page 36.1. There is a Water-retting and a Dew-retting, which last is done on a good Rawing, or after Math of a Meadow Water. Retting is accounted the finest, as indeed it is: but, as before, it must be well watch'd for 6 Hours, too long shall considerably damage it, but 24 shall spoil or rot it. H.

Page 36.3. I have recommended a Garden with this Author all along to our Farmer, than which nothing can be more pleasant, innocent and profitable: but with our Author also, that it be furnisht with things useful. H.

Page 36.4. Mast of Beach and Acorns sow'd upon the Grass in gall'd places, or in Bushes, are diligently sought after by Swine, who by rooting up the Ground, give those they leave behind the better Opportunity to fasten. Acorns are bad for Cowes, because, I suppose, the Acorn slipping into the Stomach unbroken, swells there, and will not come up to the Cud again: hence their straining as it were to vomit, and drawing her Limbs together. H.

Page 36.5 If you let him go unring'd in the Woods, ring him be sure when he goes in your Meadow or Pasture: for he will be ploughing for ground Nuts, to the great Damage of your Ground, and no great profit to himself. H.

Page 36.6. Shake time is after Harvest, when may Cattle go in the Field. H.

Page 36.8. I never knew a Hog feed on Fern or Brakes, but a Horse I have known eat young Brakes in June: If he means their Roots, a

Frost is the worst time to get at them: and I think there is little Nourishment in them at this time of Year. What is the most worthy Observation in this Stanza, is, that it was then the Custom to let their Hogs go into the Wood unring'd, where if they get no good, they do good. H.

Page 37.1. Brakes is a great part of their Firing in Norfolk, and in many Places they erect large Stacks of Brakes in their Marshes and bleak Grounds that the Cattle may shelter themselves behind them in Stormy Weather. They are very good to fence their Yards, where they night their Beasts; and if they have enough, and scarcity of Straw, they will serve very well to litter a Yard with. H.

Page 37.2. Shaken Timber is such as is full of Clefts, which unless the Sap be suckt out (as it may be by sinking it in the Water) large Timber is very subject unto, therefore the sooner saw'd the better: for when saw'd and in smaller quantities, it is not so apt, altho' not altogether free from it. Bestowing and Sticking it laying the Boards handsomely one upon another with sticks between. H.

Page 37.3. A Slab is the outermost piece the Sawyer cuts off of a piece of Timber. Saw-dust, Brick-dust, and Ashes, may make an indifferent Garden-Walk for ought I know, since in Holland I have seen pretty handsome ones made of Tanners Ouse. H.

Page 37.5. The Hook and Line is a Cord with a Hook at the end, to bind up anything with, as Wood, Hay, etc. H.

Page 37.7. So that there was a Race of Thieves in those days it seems, as well as now: but a due Execution of the Laws without Favour in the smallest Offences, I think with our Author to be the best means to prevent the greater. H.

Page 40.16. Make up your hedges with brambles and holly. "Set no bar"—put no limit: do not leave off planting quicksets while the months have an R in their names. E.D.S.

Page 41.1. Laying up, here signifies the first Plowing, for Barley it is often plow'd, so that as a Ridge-Balk in the middle, is cover'd by two opposite Furrows. This is done to rot Weeds, mellow the Earth, and to give the Water a fall from it. This he advises to be timely done, and that the Farmer be beforehand with his Ground: but as in all things there is a mean, he advises his Farmer not to be too soon. H.

Page 41.2. By Fallow, is understood a Winter-Fallow, or bringing Ground to a Barley Season (as the Country Men term it). Now if this plowing be too soon, as while the Seeds are flying, it will be the fuller of Weeds: and if too wet, the old Roots will recover themselves, and again lay hold of the Ground: Also Water running off from a new turn'd up Ground, carries with it much of its Fat and Goodness. H.

Page 41.3. Rye is sown in lighter Land than Wheat, and therefore is commonly sow'd before the Rains. When Wheat-land will not plough, which if it will not do so as to get your Seed into the Ground before Hallowmas, or All Saints, it is best to let it go till the Spring for somewhat else, for the Frosts will be with it before it can get out of the Milk. H.

Page 41.4. An Eddish, is where Corn hath grown the Year before. This is suppos'd to have weaken'd the Ground, and therefore it is proper to give it a little hold of the Ground while the Season continues yet mild, that so it may be the better able to struggle with the Rigours of the ensuing Winter. H.

Page 41.5. The Eddish in the foregoing Stanza, we may suppose was meant of what Pease or Beans had grown upon, for Wheat very often follows them: and as they are both Destroyers of Weeds, their Eddishes or Etches are very proper for it. The Pease commonly come off first, and therefore most proper: for white Wheat which is tenderer and sooner ripe than our red Wheat, Mr. Mortimer, p. 100 says, that in Hertfordshire they sometimes sow Wheat upon an Etch after Barley. H.

Page 41.6 These are the three converse ways of Sowing: for whosoever sows in Rain or over-wet Weather, shall have his Seed burst before it will sprout. He that soweth in harmes or harms way, whether of Roads, ill Neighbours, Torrents of Water, Conies, or other Vermin, can never be easy: he may lose his Crop when ready for the Barn, more likely than when in the Blade. Who soweth ill

Seed, defraudeth both himself of what he ought to make, and his Land of what it would bring forth: he hath both the Vexation of seeing his Labour come to nothing, and finding himself mock'd and pointed at in the Market. H.

Page 42.1. Of Water Furrows has been spoken before, they are commonly drawn across the Ridges in the lowest part of the Ground, so that they receive the Water from the Furrows, and convey it into some Ditch, Drain or Put: which last may be made with Success enough, where no better Conveniency is to be had, by digging past the Clay, if any, to a soft Sand, or the beginning of the Chalk-Stone, or any other Fossil capable of Clefts, through which the Water will drain. Of the use of the Bow for Children, has somewhat been said before also, only here may be added, that in our Authors time the Gun was known, altho' not in so general use as at present, and not as yet thought fit to be trusted into such Hands. The Question is, whether it were not better to re-assume the Bow, and let the Children have again the Pleasure and Profit of doing somewhat useful, than either to trust them with the mischievous Instrument which very often burns Houses, etc., sometimes destroys themselves, or else entirely give them up to their School and frivolous Sports? Physicians have long since observed, nothing is more healthful than the use of it, as opening the Breast, clearing the Lungs, etc. H.

Page 42.2. Cultivating Land, and Educating Children, is that which makes them fit for something: and again, where the best Land is neglected, it comes to very little or nothing, as some Travellers affirm of, that formerly fertile Land about Rome, which for want of due Management has render'd even the Air unwholesome.

And altho' we have no need of complaining, for want of Cultivation in this our Land, especially the more Southern part of it, yet it must be own'd, that there are still Improvements to be made: and it is great pity that a great many are so wedded to their Old Customs, as to reject Experiments, especially those that may be made with very little Cost. This is a Fault more peculiar to the English Nation, than any other that I know of: and of antient standing, at least as ancient as our Author. H.

Page 42.3. This is no doubt met with Laughter and Discouragement, until Experience shew'd who was in the Right, as it has done in many things, as Turnips, etc. since. And here I cannot but applaud Gentlemen taking some part of their Estates into their own Hands, it is to them we owe the greatest part, if not all our Improvements: for he that will venture out of the Common Road, ought to be well hors'd, and above the Bespatterings of Envious People, at least to have a Purse and publick Spirit to carry him through, for a very little Disappointment is enough to disparage a whole Undertaking. Of this we find our Author (who, what he wanted in Purse, made up in Spirit,) so sensible, that least his Design should fail, he claps Thirty Load of Muck upon his Summer Fallow he design'd for Barley: whereas upon a light Hazle Mould, Fifteen of good Horse-Dung would have done better, as we find by daily Experience. H.

"Brantham" parish, in Essex, in which Cattiwade is situated, and the place where Tusser commenced farming. The average yield of corn in his time was twenty bushels of wheat, thirty-two of barley, and forty of oats and pulse. E.D.S.

Page 42.4. There are now in use besides Folding, Horse-Dung, Cow-Dung, Marle and Burn-beating, which were known in our Authors time, Street-Earth, Mud, Chalk, Soot, Soap and Potash-Lees, Pigeons-Dung, Malt-Cummings, Lime, Sea-Coal Ashes, Raggs of divers sorts, Shavings and Shreds of Leather, Clippings of Coney-Skins, particularly the Ears: etc. Horn Shavings of divers sorts, Hoofs Sheeps Trotters, blew Clay, Urry, Sea-weed, Sea-sand, etc. All which are good in their kinds, but require Skill and Experience in the choice and use of them, wherein must be consider'd the nature of the Ground and computed the Cost, which must by no means exceed the Profit. Now altho' this last Caution may seem superfluous, as being an undoubted Axiom, yet I am so bold as to say, this is that upon which all Projects split, and therefore may very well be here remembered, especially since there goes more to such a Computation than is generally thought upon. For example, suppose I improve Land not worth a Groat an Acre to be worth Five Shillings an Acre, with very little Cost, and at the same time neglect or rob Land worth Ten Shillings an Acre, I shall run back in the latter much more than I can get in the former: or which is the same thing, if I lay out that Dung, Folding or Time, on two Crops which in another place will afford me Five. Again, if I fold more than my Sheep are able to bear, or if I keep more Sheep than I can Winter, I shall lose more by my Flock than I shall get by my Land. And here naturally enough comes in a common Error in Folding, very well observed by the Ingenious Mr. Atwell, namely, of folding a Flock by the Hurdle, or always with the same quantity of Hurdles. For suppose a square Fold contains 10 Hurdles on each side, or 60 Feet, herein may be folded 900 Sheep, at 4 square Feet to a Sheep (which altho' too little Room will serve for Explanation,) as containing 3,600 Feet. But if this Fold is removed into the Common Field, where the Ground lies in Acres and half Acres, and I am limited to a breadth, as suppose five Hurdles or 30 Feet, then the length will be 15 Hurdles or 90 Feet, and the Content of the enclosed Ground no more than 2,700, and each Sheep has no more than 3 square Feet, this being less than the other by 900 Feet, or 225 Sheep in their former space. The want of due Care in this point, over hurts and over cools the Sheep, and is the occasion of Surfeits, which commonly end in a Rot or Murrain. But to return to our Subject, as such a Computation ought to be, it ought to be in Generosity, not in Covetousness and Greediness, that is, we ought at first to be contented with a small Gain and Probability of Improvement. H.

Page 42.5. Gravel and Sand are still for Rye, not Wheat, and Pease will do tolerably well upon a stiff Land, provided they are sown with a pretty broad cast, but they delight most in a light Land that is somewhat rich. But Barley is well known to delight in a light dry Ground, such as is the black rich Mould, and will grow tollerably well in Ryelands, provided they are in heart, to which, Turneps now a days do very much contribute: so that our Author's Clay is not so proper as his rotten Land. However, if Clay be not too stiff, and brought to a good season, as the Husband-Men term it, viz. not too Cloddy, it will make pretty good shift, but in some Years is very apt to be Water-bound and Steely. It may not be improper here to add what sorts of Dung are proper for the divers sorts of Land.

Horse-Dung or Street-Muck, Lime and Chalk, are proper for stiff hungry Clays, or those commonly called cold Clays, for they mellow, fatten, and lighten them.

Marle is excellent for a light shallow Mould or sheer Ground, as Husband-Men call that Ground that loseth its Dung, one reason thereof is, because if laid pretty thick, and turn'd in pretty deep with the Plough, it forms a pan of Marle under the Soil that retain the Moisture, and the other is it that fattens and alters the Soil.

Pigeons Dung is good upon a cold Chalky Soil, and here it must not be sown too thick, for all Dungs except Marle, it is better to dung with them thin, and often, than thick and seldom.

Horn-Shavings do well upon almost any Ground, but best on light Ground: the lesser sort for Barley, and the brooder for Wheat. The like of Rags, Shreds, Clippings and Trotters, which last is by much the most lasting of this sort of Mucking.

Malt Comings is good on light Land for a single Crop: some advise it for Meadow, as also the Water of a Malt-steep, Sink or Cheese Press, in which it may be soak'd to a Consistence, and gently spread on the Ground.

Soot is well known to kill Rushes, and help cold Meadows. H.

Page 42.6. Pride and Poverty have here the same effect, that is of making the Corn lean. Pride or too much Dung (which by no means agrees with Wheat) spends its self all into Straw, and therefore where Ground is lusty, it is best to sow with a plentiful Hand: then the care of her Off-spring will keep down her Vanity. But Poverty is a more dismal Circumstance, and has no remedy, but enriching her with a Summer Fallow. H.

Page 42.7. Much Wet, especially soon after Harvest, beat down the Seeds (especially Thistle Seeds that fly in the Air) into the Ground. Hog-rooting opens the Ground to receive them. Thistles delight in dry Ground, out of which they suck a great deal of Moisture, so that they impoverish the Poor; and it is a sign that Land is Rich that is able to nourish them till they grow lusty and strong. H.

Page 43.1. It is an Old Saying, one cannot have ones Cake and eat ones Cake: for Land requires Rest and Nourishment, as well as other parts of the Creation. H.

Page 43.5. Here our Author discourseth of the goodness of the Corn, and first as it standeth on the Ground, namely, that both Straw and Ear have a proportional bigness and length, to which may be added evenness, namely, that it stand of a like thickness and height, then it is all of a piece, in least danger of lodging, and encourages both Farmer and Reaper. Then of the different sorts of Wheat, of which there are many known at present, besides those he mentions, such as Whole Straw Wheat, red Straw Wheat, Flaxen Wheat, Lammas Wheat, Chiltern Wheat, Ograve Wheat, Sarrasins Wheat: however, amongst all these, and those he mentions, the Red and White Pollard are most esteem'd: altho' they agree not with all sort of stiff Lands, or thrive alike on like Lands in all places, and therefore it is pitty that more notice is not taken of the several sorts of this Corn, and how and where they thrive best. Main Wheat weighs pretty well, but grey Wheat is, I take it, the worst, and is often ground low, and sold for better than it is at a cheaper price, to the defrauding of the Poor, and to the damage if the Market. H.

Page 43.6. The kinds of grain here enumerated are well known to wear out the land if the crops are too often repeated. He who exhausts the soil in which he has a certain interest is greedy without gain. Leases are the best preventatives of over-cropping as well as of cross-cropping. M.

Page 43.7. The meaning of this I take to be, that notwithstanding most farmers are only for Wheat, Barley, Beans and Oats with which they wear out their Land. Yet Pease and Brank, or Buck Wheat, may be a good Crop sometimes, to vary the Land and not tire it. Hence it seems, Pease were not so much used in the Field, as at present: however, now they are very considerable there, and so is Buck Wheat, which is of excellent use, as I have mention'd elsewhere, and if plow'd in the Blossom, is almost as good as a dunging. Some propose folding and feeding it on the Ground, but whether good Food for Sheep I leave to the more experienced. H.

Page 43.8. Several or enclosed Land may be used according to its strength, which in many places will hold out three Crops: but in

Common Field Land (in Most places) the Custom is for it all to lie fallow together, and that every third Year, so that the Owner of such Land must do as the rest do. H.

Page 44.1. As to his taking Wheat the third Year after Summer Fallowing, it is now out of use. I am apt to believe, by Fallowing, he means breaking up, but then Pease should have gone before Barley: however, there is a sort of Barley call'd Sprat Barley, or Battledore-Barley that will grow very well on lusty Land, but then the Ground must be fine for it, which cannot be suppos'd the first Year, so that if this was the custom then, we have got better customs since. H.

Page 44.2. The Champion way at present is most in vogue, but without doubt there may be a variation according to the divers Circumstances of places: for Example, Middlesex Men having Dung more plentiful than any part of England might afford to keep it longer, and give it more rottenness than other places, and so might fasten: and yet not overheat their Ground, so that their first Crop might be a tollerable Crop of Barley, and their next a good Crop of Wheat. But our Farmers now a days know better, than either to let their Dung waste in the Heap, or to spoil one crop to make another. H.

Page 44.4 Here he advises to be careful, near Home, of the enclosed Land, that it be not quite worn out of heart, but in time summer-fallow'd and muck'd: but for common Field-land, and what lies remote, he looks upon it as no great matter how near it be worn: however, he recommends the comforting it with Favour and Skill. By Favour may be understood laying it down: but what his Skill was, he has left unfolded, for I took it in those days there was nothing for it but folding: Perhaps there might be some other ways, which were Secrets: such as sowing Tares, or Brank, and ploughing them in, using Rags, etc. I am sure we have a ready remedy at hand at present, were it not for spite, which is by Turneping them, and feeding Sheep or Neat Cattle upon them, by making a Fold of Hurdles: but this the Owner of the Sheep-Walk commonly eats up before it can come to any Maturity. H.

Page 44.5. Codware, such as Beans and Pease, are observed to be no great Peelers of Ground: Beans delight in a stiff Land, and Pease in a lighter Mould: of which before. H.

Page 44.6. The Case is very much altered, for they are as much crav'd as any other Commodity, and the Husbandman may make much more of them in Money, than in Hogs-flesh. One Reason is (I believe) because there are not near so great a plenty of Acorns as formerly, with which the poor Man used to fat his Hog: and altho' his Hog already stands him in more than he is worth, he must not lose the Feasting and Joy, this Creature is like to afford him and his Household, for fear of a little farther Loss: However, hitherto what his Hog has cost him, has gone away insensibly, and will do him no more good in Pork, or Bacon, than if it had been in Ale. H.

Page 44.7. Two Crops of a Fallow is pretty well, but I think one Crop and a Fallow but very poor doings: however, it is better to bestow Labour on the Ground, than to lose it in the Crop: Now between a good Crop and a bad one, there is little difference in the Ploughing, Seed and Inning: but there is a vast deal in what they make at the Market: and the Labour of Fallowing is better laid out at Home, than lost at the Market. Here it is again observable, that Pease were lookt upon but as an indifferent Crop. H.

Page 44.8. Pease are no impoverishers, but rather improvers of Ground, so that if you have a Fallow after a good Crop of Pease, he supposes the Ground still in heart enough to bear a Crop of Wheat: for too much Dung, or too much Water, are bad for Wheat: we have observed elsewhere. H.

Page 45.1. Because these are sown in the Spring, when the Water is going or gone off: and besides, these are not so apt to burst as Wheat: Bullimong has been elsewhere explain'd: it is a mixture of Oats, Pease and Vetches. H.

Page 45.2. This is for the raising a Wood, which will very well bear with a Crop of Rye, taken off the first Year: for both Acorns and Haws, being very slow in coming up, will not be very far above the Ground at Harvest: But then they must be well fenced from Cattle and Cony, the first two Years after, as also very clean weeded, after which they will require little tending, except the Fences. H.

Page 45.3 The Reading and Hasting, are best sown at this time of the Year, which if they take good Root before the cold Weather comes, with some Care and Favour of the Weather, may live until the Spring: but they have a great many Hazards to run from Black Frosts, etc. H.

Page 45.4. Some advise the twisting the Seeds in a Hay-band, and so bury them shallowly in Rows. Be it how it will, they must be fenced in, and then it will be found that a new Bank with Quicksets, is as cheap. Of the raising of Haws and Sloes, I spake before, and I believe the Bramble may be rais'd the same way: namely, by burying the ripe Berries during the Winter until their Seeds chitt, and then sowing them. I am also of Opinion, that a Bramble may be planted with good Advantage, as Vines in a Vineyard, and with good pruning and ordering, may be brought to ripen altogether: which if once they are, they will be of excellent use. H.

Page 45.5. If fed under the Tree whilst green, and in moist Weather, or if fed with any thing that is too cold and moist, as Gardning Pease that have taken wet, etc. Hogs are very apt to swell under the Throat to a prodigious bigness, which if not taken care of in time, choaks them. The best Remedy is giving them their Wash hot, and if ripe, cut open the Swelling, and the Matter will spurt out a great way. This I have known done with success, but I take it the best way, is to pierce it an Inch or more deep with a red hot Iron. H.

Page 45.6. This is a plain matter of Fact in all Edibles. H.

Page 45.7. On the contrary, it is a horrid thing for Farmers and others, to sell that which dies of its self to poor hungry Wretches, who as greedily eat it, and suck in the Venom, which is very frequently done. H.

Page 45.8. Whether Measled Bacon be infectious or not, I cannot tell: it commonly happens from the Hogs want of Water: however, if the Flemming delights in it, and buies it knowing it to be such, it is a pity to eat any from him. H.

On that part of the coast where our author lived many Flemings had settled. It is evident that they were not delicate in the choice of their food. M.

Page 46.1. This is what is call'd Light-Fire in Norfolk, and serves excellently well for those uses. Seething of Grains is no bad Husbandry, especially at this time of the Year: for altho' little is got out of them, the Heat is very comfortable to the Hog. H.

Page 46.2 By Sweating in the Mow it has contracted a Thirst, which by the Air is cooled, and the Spaces plumped, so that the Flower separates much better from the Bran when ground. H.

Page 46.3. They are an excellent and cheap remedy for laxity of the bowels, in men and cattle, if judiciously used. M.
They are best bak'd gently in an Oven. H.

Page 46.5. This Medicine retains its Credit to this day, and it is much to be admir'd that we should give so great a Price for Lemons and Limes from abroad, and despise the Crab, of which Verjuice is made; an Acid no less pleasant, and more improvable, than what comes from any of them, only because we may have plenty of them at Home. H.

Page 47.2 As Tusser is pretty correct in his rhymes, he probably wrote *beasty* originally. M.

Page 49.1. In some Countries they kill Pork all the Year long (as at the Bath, etc.) with Success enough. However, this time of the Year affords Off-Corn to keep up what they got in Harvest, and Beans and Pease are now most plentiful; the Season of the Year also, by reason of its Coolness, is most proper for Fatting. H.

Page 49.2. Dredge is a mixture of Oates and Barley, and at present used very seldom in Malting, as not working kindly together, especially when they are to be wrought for increase of a Bushel in a Seam or Quarter, as our Author here intimates. H.

Page 49.3. This is meant of cleaning of Barley, which for malting need not be so clean as for Seed; for the light Corn may be skim'd off at the Cistern, and if the Cockle be left in, it will work, and some say, make the Drink the stronger; but whatever ease this may be to the Farmer, the Malt-Man, if he be wise, will make him pay for in the price of his Commodity. H.

Page 49.4. Stover is Food, and in Winter dry, and lean Cattle will make very good shift with Barley or Oat-Straw. It is best to feed them from the Thresher, both, because it is then most juicy, and to avoid pestring the Barn. H.

Page 49.5. Wheat is well known to work better in grinding and baking, after it has undergone a natural heat in the rick or mow. Wheat that is threshed early keeps with some difficulty. M.

There are many way mention'd by ingenious Authors to preserve Wheat in Granaries, as mixing Beans amongst it, Pipes to go through it with Air-holes, the running it through holes like Sand in an Hour-Glass, from one Floor to another. But all (if practicable) come short of keeping it in the Sheaf, from whence it goes to Market in its true Beauty. Next to the Sheaf is the Shovel, namely, by frequent turning, and thus it is preserv'd in Holland and Dantzick, notwithstanding the moistness in the Air. H.

Page 49.6. Feathers are very noisom in the Food of Cattle, especially Horses, who chew all down. To prevent Fustiness, the best way is to eat it off as new as you can, for Chaff is very apt to attract Moisture, and Moisture is the occasion of Mouldiness or Fustiness. H.

Page 50.1. A good Crop of Pease to be sold in the Shell, is worth any Man's looking after: and if they are sown now, unless a black Frost come, they are like to be very early. If they are nipt, it is worth while to sow them again, or drill where there is wanting, for a Peck in the Shell is seldom more than a Quart, which at 6d. the Peck, comes to 16d. the Bushel: and the Hawm of Foreward gather'd, Pease is little worse than Hay, besides there is time to have a good Crop of Turneps the same Year. H.

Page 50.3. This is all very plain, and generally understood, and of which much has been said before: there remains only to consider, what share the Poor have in the Farmer's Poultry, which I suppose was no other, than that the Thresher and other Days-Men had the running of a Pig or two in the Farmer's Yard, which if the Farmer was overstock'd himself, it is likely they could no longer have. This, as well as many other parts of the Old English Hospitality, is very much disus'd, and perhaps not without very good reason. H.

Page 50.4. This we see daily verify'd, namely, poor Wretches that cannot maintain their Families, must have their Dog or two after them, tho' they know they are maintain'd to the prejudice of their Betters. It springs from a sort of beggarly Pride, or desire to live at the Publick Charge, and I think a Man ought to be call'd as much to account how his Dogs live, as how he lives himself. H.

Page 50.5. Smoak dry'd Meat was in much more Request formerly than it is now a days. It is true, Smoak gives a Firmness and Durableness which makes it fit for Exportation, etc., as well as a Gratefulness of Taste. But then it is hard of Digestion, and liable to much Waste: and therefore justly left off in many Places, and Pickle prefer'd to it, which both better preserves the Meat in its natural Taste and Sweetness, and makes it spend with less Waste, it saves Salt also. H.

Page 50.6. St. Edmund is on the 20th of November, at which time it may be very proper to set Garlick and Beans: but why the Moon in the Wane? I cannot tell unless it be that he thought in the Wane, the Weather grows warmer and warmer until the new, because the Moon is then continually approaching the Sun. H.

Page 50.7. That is, the Plough-man, Horse-keeper, etc., who commonly like not this sort of Work, and if they are not watch'd, will leave more in the Straw than the Work comes to.

Page 50.8. That is, be as much with your Servants of all sorts as possibly you can: for the Eye of the Master makes not only the Horse fat, but his Work good, and the Servant careful. H.

"Plaie tapple up taile", a cant expression, meaning to tumble head over heels. E.D.S.

Page 51.1. This of the Bottle, I remember I heard a Farmer say, he once found out, and there are still too many pilfering Rascals of this sort. But then on the other hand, how can a Farmer expect a poor Man should live upon such small Wages that they sometimes run them down to? I have known when, and where a Thresher could not get his 4d. a day, and had at the same time a Wife and Children to maintain: if this Fellow had been sent for a Soldier, he had got by it. So that altho' the Law is defective in this point, yet, methinks, Conscience should dictate to us, that we ought not to desire any Man's Work for less than he can live upon, any more than he should take from us more than he bargain'd for. H.

Page 51.2. If there is House-room and a Market near, Straw, especially wheat and Rye-Straw, may very well be laid up: but if no good Sale for it, after you have sav'd what you think fitting for Thatch, and fodder'd your Cattle, and litter'd your Horses, the rest may lie in the open Yard, for the Cattle to tread into Dung, which is the practice now a days, so that our Farmers are not so afraid of noying their Doors it seems as formerly, and that not without good reason. H.

Page 51.3. By which Means you may see if your Corn yields alike. H.

By making up the floor weekly less waste is committed, and less temptation given to those who are disposed to be dishonest. M.

Page 51.4. It may lie much better as I said before, up and down the Yard, especially in the lower parts of it, where the Cattle go, for Straw retains Moisture, and as it becomes rotten and full of Dung, it may be cast up in heaps and carried away, too much Water weakens it, unless the Water be the Fat of the Yard. H.

Page 51.5. By Head-lond, I take it, is here meant such Ground in Common Field-land, which the whole Shot (or parcel of Land belonging to many Men against which it lies) turn upon. This cannot be sown until all the rest have done, and perhaps in our Authors time, was seldom sown at all, and it is its new breaking up which he alludes unto, when he advises to cast it up in Hillocks to rot, meaning the Grass-swerd. H.

Page 51.6. Garden Trenching is excellent good for Carrots and Parsnips, and indeed for any thing: it is the best way of Mucking. H.

Page 51.7. Humane Ordure has for a long time been thought unfit for Land, as being too fiery: but this Heat may easily be allay'd with Straw, Fern, Earth, or any Vegetables, to give it a Fermentation, and then it is the greatest Improver of any Dung whatsoever: Mr. Mortimer, p. 23, says, it sels in Foreign parts at a much greater rate than any other Manure. H.

Page 51.8. Our Author here mentions the Mischiefs arising by Soot, but I believe was ignorant of its Benefits. In short, it is now found to be one of the greatest Improvers of cold Clay-land, whether in Corn or Grass, that the World affords: and particularly destroys the Moss in Grasslands, for which Disease, it may be justly esteem'd a Specifick. H.

Page 52.1. Trees and Plants particularly the Gooseberry and Hop, are only lousy in dry Seasons. The Nits which we frequently see upon the Shoulders and Flank of a Horse, are blown by a sort of Fly, very like a Bee: which I believe are gone before this time of the Year, so that I am not clear in our Author's Observation. Poor Horses will be lousy whether the Season be dry or wet. H.

Page 52.2. Compass is Dung, of which the Yard should often be clean'd, that the more may be made: and whatsoever a Lady may think, a Farmer thinks heaps of Dung a very good Ornament to his Dwelling. H.

Page 52.3. That is, of the tops or parings of the Mold-hills, although with as good Success they may (after 3 or 4 Spits of Earth are thrown out) be laid down again. They ought, however, to be laid for some time open, that the Wet may destroy the remaining Pismires, however, for the sake of that industrious Creature, let me add, that altho' they are an Annoyance, and the Farmer may improve his Ground by destroying them, yet where they are in Pasture to be fed, they do least harm: and the Hills are an excellent Shelter for Lambs unless they stand too thick. H.

Ant-hills, as increasing the vegetable surface, have been considered by slovenly farmers, who wanted an excuse, as an advantage, but their only utility seems to be in affording shelter for lambs, and it is not necessary that they should be very numerous for this purpose. M.

Page 55.1. Frosty Weather is best for the Dung-Cart, but when that is done, our Country-Man may employ his Servants with his Beetle and Wedges, much better than by letting them hover over his Fire. H.

Page 55.2 A Grindstone is very necessary about a Farm-House: It keeps the Servants from gadding to the Smith's Shop upon every small occasion, which with the Mill is the Seat of News. H.

Page 55.3. This is meant of Foddering in the Pasture Grounds, wherein care ought to be taken that too many be not fed together, for the Old will be apt to hunge or gore the Younger. H.

Page 55.4. The Rack must be so set, as that the Young may reach it, and easily run under it: by which means they escape the Hunges of the greater Cattle, and at last get a quiet Feeding place. Shelter from the North and East Winds, is as good to Cattle as half their Food. H.

Page 55.5 The Housing of Cows, as frequently used in Hertfordshire, is certainly the best way, both for safety and husbanding their Food: but I think there is little to be got, or sav'd, by housing other Neat Cattle, unless Stall-fed Oxen for the Butcher. H.

On the flat and stormy sea-coast where our author lived this caution is highly proper. M.

Page 55.6. By this Stanza it seems as though he recommended the Housing of Weanlings, which perhaps may be worth while (if Cow Calves) for the first Winter, but I do not remember to have seen it practised. H.

Page 56.1. It often requires the Strength of a Man to lift up poor Cattle, who sometimes cannot rise when they are laid, especially in Snows and cold Seasons: and therefore I suppose this is meant of Foddering in the Field, as well as feeding in the House, where without doubt there may sometimes be need of Help also. Turnep feeding, as used in Norfolk, requires a constant Attendance, and also a strong Hand: It is frequent for the Cattle to be almost choak'd by a piece of a Turnep, lying a cross in their Gullet: to which end, the Tender has a Rope of a pretty large size always at hand, tufted at one end: this he supples with Butter, and by thrusting it down the Beasts Throat, pushes the Turnep into its Stomach. H.

Page 56.2. Rie Straw is of all Food the poorest, and indeed seldom used to that purpose: however, our Author's meaning is, that the worst shall be used first: but then you must begin in very cold Weather. H.

Page 56.3. The Reason of this is double: namely, not only for the thing, but to shew where he has been, and that he may not pretend to have been in the Woods, when he has been at the Ale-house. Bathing at the Fire, as it is commonly called when the Wood is yet unseasoned, sets it to what purpose you think fit. H.

Page 56.4. The highest Spring Tides are not only in March and September, but when the Wind has held for some time, before the Full or Change, against their coming in: and therefore the third day is commonly the highest Tide: for although the Wind does not always hold against it, yet the Current of the River where it sets in does, which amasses the Waters to a superiour Strength. H.

Page 56.5. At present this is not so needful an Instruction as formerly, because Farmers either find it not worth their while, or are not willing to keep any Lenten Days. I have spoke of this in former Months, as a very great Neglect of the Blessings of God, and therefore shall say no more at present than this, that if we despise the Product of the Sea, a Neighbouring Nation knows how to make use of it, to our Eternal Shame and Reproach. H.

Page 56.6. Wind-dry'd is the best drying of Fish, especially dry cold Winds. H.

Page 56.7. About Christmas, that is when the Sun is in the Winter Solstice, the Sap is thickest: and consequently the Tree is less sensible of its remove, being as it were asleep. H.

Page 56.8. Some set between every Apple-tree a Cherry-tree, which at 12 Years growth is cut down, and by that time the Apple-Trees are come to their due spreading. It is very material upon transplanting, to plant exactly in the same Situation, in respect to East, West, North and South, as it stood before, especially when the Trees have attain'd to any Grandeur. H.

Page 57.1. That is, when once planted, afterward neglected. There is one thing needful in an Orchard, the want of which is the occasion of the most part of our bad and unsavoury Fruit: namely, taking Trees upon the Gardiners Words, or, because they are of a good kind, in one place taking it for granted they must be so in all, whereas they will not only thrive so well in one place as another, but degenerate, and become worse: and therefore as soon as your Trees begin to bear, if the Fruit please you not, extirpate them, and plant others in their room. H.

Page 57.2. About Christmas is a very proper time to bleed Horses in, for then they are commonly at House, then Spring comes on, the Sun being now coming back from the Winter Solstice, and there are three or four days of rest, and if it be upon St. Stevens day, it is not the worse, seeing there are with it three days of rest, or at least two. H.

Page 57.3. Pease-hawm, or Straw, that comes from such Pease as have been gather'd in the Shell, is what is here meant by green Peason, and is apt to gripe Horses who will eat it very greedily. The Remedy is scalded Bran. H.

Page 57.4. The mean to be sure is the best, Horses and all working Cattle ought no more to be pamper'd than under-fed, especially with Corn. H.

Page 57.5. Fitches or Vetches are of divers sorts, of which before, but since our Authors time, several new Grasses have been found out, which supply the same Defect. Those which are most in Request at present, are, Clove, Ray-Grass, Nonesuch, and St. Foin. H.

Page 57.6. The best come out first and easiest, and therefore most proper for Seed: what is left in the Straw does the Horses good: but neat Cattle, and what chews the Cud, hard Corn is lost upon. H.

Page 57.8. It is very comly and looks like delighting in Home, when a Garden is well look't after at a Farm-House: not with fine Walks and Winter-Greens, but things useful. H.

Page 58.1. Or ye may spread a little Honey on a Board, which I take to be much better. It is true, some Years this may not be amiss, especially about this Month: but if the Fault be in the Weakness of the Stock, not in the Wetness of the past Summer, they are not worth feeding. H.

Page 58.2. Camping is Foot-ball playing, at which they are very dextrous in Norfolk; and so many People running up and down a piece of Ground, without doubt evens and saddens it, so that the Root of the Grass lies firm: altho' at the same time the Horse-Men do it not much good, especially if it be somewhat low and moist. The trampling of so many People drive also the Mole away. H.

Page 61.4. "Ever among"—constantly, frequently: an expression of frequent occurrence in Early English. E.D.S.

Page 67.4. There is some confusion here; probably we should read, 'and *flies* from sinne,' *etc*. M.

Page 70.3. "Yet Michel cries" *etc*.—to delay the operation of cutting, and therefore the cries of the animals, till Michaelmas, will have the effect of getting them into such condition as better to please the butcher's eyes. E.D.S.

Page 72. "Perareplums"—variety of plum either lost or unknown, if not a misprint. E.D.S.

Page 73.1. The Author liv'd the greatest part of his time in Norfolk, Suffolk, and Essex: in the two former there is much Cattel reared at present, the latter is much altered from what they did formerly, because of the Profit they make by suckling Calves, and housing of Lambs, and the taking in of Commons. H.

Page 73.2. He advises the Farmer to kill as good as he sells, perhaps to credit him when his Chapmen come to buy: else Experience tells us, if he eats not his old Ewes and such ordinary Meat at home, he will get but little for them of the Butcher; for best is best cheap only when 'tis bought. So as to his rearing of a Pig, if it be in a poor Man's House, or one who buys all with the Penny, his Souse may be sweet, but his Bacon shall be dear. H.

Page 73.3. Broath is still us'd in some Farm Houses for Supper Meat, and roast Meat look'd upon as very ill Husbandry. But if the Farmer hath latter Pigs, Calves, or Lambs, which the longer they are kept will be the worse, he may eat them or sell them whilst they are good, and for want of Broth make shift with better Liquor. H.

Page 73.4. This holds good still: foyzon is Winter Food. H.

Page 73.5. Carry out dung in frost, so that the carts may not cut the roads or the grass. M.

Pease boyling or not boyling is one of the Farmers occult Qualities, but fresh, and next to it, well dunged Grounds are observed to produce the best Boylers, perhaps because they retain most moisture. H.

Page 73.6. By Experience Garden Quicksets are found to be the best, they as well as others ought to be set in new thrown up Earth, and weeded the first two Years, which is done with much Ease: The Gardens are preferr'd because they are all of an Age. Poplar and Willow Stakes will grow in a Clay or any kind Mold, but they assuredly dye as soon as they touch the Gravel, perhaps it is too dry to afford them Nourishment. H.

Page 74.1. The common time of ending their Slaught (or Slaughter as the Warreners term it) is Candlemas, altho' they often leave off sooner, as in Case of a mild Winter: the Flesh is red and unsavory soon after Christmas. The use of Pigeons Dung is now better known than in our Author's time. As to cleaning a Pigeon House, some with very good Reason defer the taking away the Heaps of Dung that lye before the Pigeon Holes, because they are a good defensative against the ensuing Cold, and preserve the Eggs, and Pigeons of the first Brood. H.

Page 74.2. Since the Author's time there are many better sorts of Pease to be set at this time, but the most forward Pea is the Rogue, they are pick'd from the Hasting and Hotspur, and are of late had in great Reputation. H.

Page 74.3. This happens according to the Chapman's want of Money. H.

Page 74.4. It seems the Servants Fire was biggest then, and so it will be still if care be not taken: however if they have none but what they must cleave the Moment they want it, it will somewhat lessen the Wast. H.

Page 74.6. Since the use of Turneps Cattle need not be hard put to it, in snowy Weather as formerly, but still they are in hard Frosts, and then nothing agrees so well with them as browze: the like of Deer. Conies will grow fat upon browze if they have but enough. H.

Page 74.7. Edder is such fence Wood as is commonly put upon the top of Fences, and binds or interweaves each other: Stakes and their use need no Explanation. H.

Page 74.8. This is more proper in Underwood than Pollards, at least more in use at present: few Pollards perish for want of it, but Runt-wood [stumps of underwood] will. H.

Page 75.1. In Gauls of Underwood this may be done with some Advantage (Gauls are void Spaces in Coppices which serve for nothing but to entice the Cattel into it, to its great Damage) and then the best way is to let your Loppings lye some time before they are fetch'd away: but there are much better ways than this, particularly by sowing Acorns on the Grass, which will take root and turn to better Account. H.

Page 75.2. It is not enough to give the Hewers in charge that they cast out every thing to the best Advantage, but they must also be watch'd and encourag'd, by giving a Reward for every hundred of Stakes, bundle of Prick-wood, or score for Poles, etc. H.

Page 75.3. It is certain that having a thing at hand when wanted, and seeking it or borrowing it, it is in a greater Proportion than as one to two. H.

Page 75.4. Prime Grass appears commonly in woody moist Grounds, on Hedge Banks, and is so called from its earliness: when Cattle have tasted this they begin to loath their dry Food. It is often sprung before Candlemas, for the Spring may properly be said to begin from the Sun's returning from the Tropick of Capricorn. H.

Page 75.5. Verjuice is well known to be the Juice of Crabs, but it is not so much taken notice of, that for Strength and Flavour it comes little short, if not exceeds Limejuice. H.

Page 75.6. This Remedy still is in Practice, how reasonable let the Learned discuss: however, by Experience we see, that the first Indication of corrupt Blood is from the staring Hairs on the Tail near the Rump. Some instead of Soot and Garlick put a Dock Root, or the Root of Bears Foot, which they call a Gargat Root, others flay the Dewlaps to the very Shoulders. H.

Page 75.7. Large Ant-Hills is much the best Shelter for Ewes and Lambs: a Broom Close is also good: but the worst, to be sure, is Bushes, for as they grow weak their Wool is dryer, and more apt to flake of. H.

Page 75.8. Good fence Wood in a Farm, and enough, is half a Crop. H.

Page 76.1. This is understood of Hedge Greens: that is, in every arable Close, there is a space next the Hedge, of a Rod or more in breadth, left for Pasture, this ought to be kept clean from Bushes: which if it is not, it is natural to the next Neighbour, when he mends his Hedge, to cut them to his Advantage—Belive signifies in the Night, which is more put in for Rhime sake that the Neighbour should be suppos'd to work in the Night. H.

Page 76.2. This is when you rid it of Bushes or Ant Hills, but when you rid Ant Hills it is best to throw out a pretty deal of Earth, and return your Turf so as that it may lye beneath the Surface, as the bottom of a Dish to the brims, for then it will gather the Water, and kill the remains of the Ants. H.

Page 76.3. This I take to be meant still of Hedge Greens, which after fencing have a pretty deal of dry Wood or Stubbings left on them, which the Farmer ought to carry home for his use. Hugh Prowler is our Author's Name for a Night-walker, for whom he would have nothing left: however, we may suppose they suffer'd the Poor to glean Chips, and small bits after the Cart.—There are a sort of Wheels call'd dredge Wheels, now in use, with the help of which a Load may be carried through a Meadow, altho' it be not a Frost.—If the Land be stony, the Plough is apt to turn Stones upon the Green, which must be pick'd off again. H.

Page 76.4. A Barth is commonly a place near the Farm House well sheltred, where the Ewes and Lambs are brought in for warmth, and the Farmers Eye against these six Enemies. H.

Page 76.5. By dainty I take it is here meant likely or thriving, such a one as will soon require more Milk than his old Dam can afford him, and therefore most proper for the Knife whilst he is good, but since the housing of Lambs this Rule may be varyed——. There is little Ewe Milk used in England, but where they do, it is proper to keep the Lamb so long by the Dam's side until she has Plenty of Food: to be sure she will give all she can down to her Lamb, and when her Food is plentiful she must do the same to the Pail. H.

Page 76.6. "*Peccantem*" should be *peccavi*, which is the reading of the editions of 1573, 1585, 1595. E.D.S.

That they may not sing *peccavi* they put them not to Ram until a Fortnight after Michaelmas, so that they fall about the beginning of April or latter end of March. H.

In some part of Norfolk and Lincolnshire they will keep none but Twinlins, but then it is in rich Land, as Mershland and Holland. H. Twin lambs, though I know not on what ground, are supposed to perpetuate their prolific quality, and are kept for breeders. M.

Page 76.7 Forward Calves after Christmas, are to be sure the best to rear, as having a long Summer before them. The Prime is the first three Days after the New Moon or Change, but for what reason those who come within that time must be killed, I leave to the more experienc'd: 'tis true, those Days are most subject to Rain. H.

Page 76.8. At present we rarely we wean under twelve Weeks, but in Lancashire such as are design'd for Bulls suck much longer. H.

Page 77.1. They must be taught to eat Hay before they are wean'd, which that Calf that takes to first may be said to teach the other: the Hay is given them stuck in cleft Sticks, and must be of the finest. When they ail any thing they are not so skittish as when well, and therefore will endure and be us'd to stroaking better than at any other time, or perhaps it gives them some ease, which they remember. H.

Page 77.2. For rearing, if the Calf be a Fortnight old and the Lamb five Days it will do as well. H.

Page 77.3. Because the Pig farrowed now will be Pork at Michaelmas, or Bacon at Christmas next, and Wash becomes plentiful by the time they are weaned. H.

Page 77.4. It is likely that the strongest Pigs get foremost, and the foremost teats are generally suck'd lankest, and consequently give most Milk. H.

Page 77.5. To be sure they will grow apace, and the Sows will not go to Boar until the Spring following, so that they will have time for growing too. H.

Page 77.6. Gelding is still done under the Dam, but spading is more frequently deferred, and that with Success enough. H.

Page 77.7. This agrees with our present Practice: the best way of gelding Colts is with an actual Cautery. H.

Page 77.8. It is difficult Work, and requires a skilful Hand, but may be deferr'd longer: it is not much in use because of the many Disasters attending it. H.

Page 78.1. It is a creditable and joyful Sight to see a fair large Breed on a Farm, but then it ought to be proportional to what the Farm will carry off, not Lincolnshire Sheep on Bansted Downs, or Lancashire Cattel in Northumberland. H.

Page 78.2. This is to be understood of Cows kept in good Pasture, not the poor Man's Cow which runs upon the Common, which besides his loss of Time after her, seldom pays her wintering.—A Sow may be as profitable as a Cow, provided her Pigs are sold for Roasters, and have a good Market: neither must their Food be bought by the Penny, but where Sow and Cow are kept together. H.

Page 78.4. The Author was for some time a tithing Man, and it is likely he found many Farmers grudge at so considerable an out-let of their Crop, for it is indeed little less than a Sixth: but if they are convinc'd it is the Owner of the Land, and not they that pay it, they may be more easy. H.

Page 78.5. In trenching bury no Mallow, Nettle-dock, or Briony Roots. H.

Page 78.6. Hops love their Head warm and Feet moist, however not too moist, but a pure light rich Mould is best. H.

Page 78.7. Quicksetted Arbors are now out of use, as agreeing very ill with the Ladies Muslins: howsoever it holds in Espalliers, and all other Pole-work, not to pole or wattle until there is a growth to menage: Wattles are Wood slit, such as in some places Gates are made of: in their room we more neatly at present use slit Deal. H.

Page 78.8. This is a celebrated Stanza, but, I doubt, seldom practic'd, yet perhaps both may be done to Advantage: for such early sown Oats it is likely may be clearer of Weeds: & if I buy my Hay in May, that is, before my Chapman knows what Quantity he shall have, he is rul'd by his Necessity for some ready Mony in Hand. H.

The whole stanza is somewhat enigmatical. The standard editions uniformly read, "by the hay" *etc.*, but the more modern have, "buy thee hay" *etc.* M.

Page 79.1. This is understood as the former, of breaking of Lay, which, if troubled with Roots or Gammock, a Servant is very well bestowed to be ready to clear the Plough before all Flies. H.

Page 79.2. If it be Grass, break it up as soon as you have mow'd it, or fed it down: then instead of your after math, or latter feed, you will have a Crop of Corn the next Year. H.

Page 79.3. Barley is now very rarely, if at all, sown on lay Land, the Fallow he speaks of I take to be the second ploughing for Barley, which every one must be guided in, according to his Circumstance of Team and Quality of Land. H.

Page 79.4. Barley Ground ought to be as fine as an Ash-heap, as the Country People say, and if you find it rich enough for a Crop of Barley never Oat it, for that may come after. H.

Page 79.5. Where the Mould is shallow, and the Ground dry, it is not good to begin with Oats, but where the Ground is over rich it fines and sweetens it. It is a common thing in the Isle of Ely, and other Parts where the Ground is over-rank and course in Grass, to take off a Crop of Oats and sometimes two, and then lay it down again, and the Ground will be much the finer, and the Grass sweeter. H.

Page 79.6. If Ground could be worn quite out of Heart, a Crop may be as well expected from a Stone: but when it runs to nothing but

Carlak or wild Oats, or if clean, will not afford three times your Seed, it is then worn to the Proof, and does require rest, folding, or dunging. H.

Page 81.10. "Plash"—to pleach down a hedge over the burrows; "set"—plant over the place where the burrows are, not to stop the rabbits from coming out, but to give them a means of escape from the dogs who might otherwise *snap* them up before they reach their holes. E.D.S.

Page 83.1. It is not usual, at present, to let the Dung heaps lye a Month, or any longer Time upon the Ground before it is spread than Conveniency and Opportunity requires: it is also proper, if the Dung-heaps have stood any time, to take some of the Earth on which they have stood, and spread it abroad as Dung: and when all that is done, when your Crop comes up, you may easily see where they have been, they will so ranken the Ground. So that I take it, our Author here means a Field Dunghil, which indeed ought to stand some time: but then this is not the proper Season to make them, at least as Husbandry is now practis'd. H.

Page 83.2. The Furrow is the barennest part, as being the lowest (if the Soil be shallow) and (to be sure) the heat and moisture of the Dung-heaps will fatten it about equal to the rest. Let not your Dung however stand too long unspread, for fear some of its Fat sink out of your reach. H.

Page 83.3. The Stubble had better have been ploughed in before, especially if it be Wheat or Rye Stubble. Beans delight in a stiff Mould, and are no Peelers, for they fetch their Nourishment deep. Pease, and Fitches or Tares, delight in lighter Mould, and are great Destroyers of Weeds, and for that reason are also no Peelers. There is now a Winter Fitch or Tare much in request, which ripens much sooner than usual, because of its early sowing, and consequently remedies the greatest Inconvenience that attends this Pulse, which requires more time than Pease. St. Gregory's Day is the 12th of March, before which white Pease are now frequently sown: but grey Pease always are sown soonest. H.

Page 83.4. Planetary Influence, especially that of the Moon, has commonly very much attributed to it in rural Affairs, perhaps sometimes too much: however, it must be granted the Moon is an excellent Clock, and if not the Cause of many surprizing Accidents, gives a just Indication of them, whereof this of Pease and Beans may be one Instance: for Pease and Beans sown, during the Increase, do run more to Hawm or Straw, and during the Declension more to Cod, according to the common Consent of Country Men. And I must own I have experienc'd it, but I will not aver it so as that it is not liable to Exceptions. H.

Page 83.5. If you don't, the Vermin, as Rooks, Pigeons, etc., are sure to have a good share of them: as they will (unless you watch them) if you do: for the Rook will watch them when they first begin to peep out of the Ground, and time it very exactly. The Pigeon always begins where he left of, and will (if he may) go over the whole, and make of it an entire Piece. Add to this, that in some measure both these are lawless Thieves, and therefore must be prevented by hiding and scaring only. The reason why unharrowed Beans set in Clay are apt to dye, is because the Wet fills the Holes and rots them. H.

Page 83.6. This is called sowing under Furrow, being sowed on the Land just before the second ploughing, which if neatly done, lays them in rows just as if they had been drill'd: And here falls in another Reason why Pease and Beans ought to be soon harrowed in, because if they lye until they are swell'd the Horse-footing is apt to endamage them. H.

Page 84.1. This particularly regards Field Land: for in our Author's time Enclosures were not so frequent as now. There every body ought to consult his Neighbour's Interest as well as his own: for it is hard, that for my Negligence, in not sowing timely, my Neighbours Swine and Cattle should lose the Benefit of the Field, and that the Sheep should sweep it before it is half fed: which, by the way, is no Benefit to the Sheep neither (as some Shepherds well observe). Take Care also not to sow Winter Corn upon such Headlands as your Neighbour must necessarily turn his Plough upon. Also in enclos'd Land be not behind your Neighbour, if possible, especially if the Fence be yours, lest you be forced to make up your Fence when the

Ground is too dry, and you have no time to spare from your Harvest. H.

Page 84.2. There is nothing got by under-feeding working Cattle, nor is any thing got by over-feeding them: Their Food is to be proportioned to their kind of Work: for Cart-Horses and Saddle-Horses may be very well look'd upon, as of two kinds, the swifter their Motion, the lighter and more spirituous ought to be their Food. Oxen will work very well with good Hay: Cart-Horses require some Provender, and will do very well with Chaff and Oates: the Saddle-horse requires good Oats and Beans: and these deserve their Food no otherwise than as they pay for it with their Labour. H.

Page 84.3. There were such poor People in our Author's time, it seems, and so there are now a sort of People who take a World of Pains, and do a great deal of Labour to be poor, wretchedly poor: What Necessity, and want of ready Money may plead for them, I cannot tell, but this is certain, that whosoever loses his Season for sowing must expect almost a Miracle in his Favour, or he must compute short of a Crop. Now the Question is, whether he has not as good rely upon Providence at first, before he provokes the Almighty H.

Page 84.4. Well fed Cattle will do their Work merrily, and thrive upon it: and it is evident that the Work of a Beast is equal at least to four times his Food. What a silly Covetousness is it then for Men to lose a third Part of the Work to save a fourth Part of the Provender, for more cannot well be pinch'd: besides the Danger of losing the Cattle. Yet such People as these there are in the World, and a great many too. H.

Page 84.5. In the last Month I recommended Garden Quicksets as the best: next to them are the smallest, and such as have the Roots fine threaded: by no means meddle with stubbed ones, for they are but part of old Bushes. The manner of raising Garden ones take as follows, At Michaelmas get a Quantity of Hawes, and bury them in an indifferent Mould, not too rich, until the April following; then you shall find them lying in a black Lump, the most Part of them chitted or sprouted: separate them gently from each other, mixing them with some fine Mould: then sow them on a well prepared Bed of good Earth, sift over them Mould about a Finger's Breadth thick: weed them carefully the first Year, as often as you see any Weeds amongst them: the second Year, at least four times: and the third Year, at Michaelmas, you have as good a Crop as your Garden can produce. I advise that the Mould wherein they are sown be very good, not barenner than what they are to be transplanted in, as some teach: for every thing has its Infancy and time of Tenderness, in which it must be tenderly used, and have fitting Nourishment. The jolly Lad that has been well fed in his Cradle, is certainly healthier, and able to endure more Hardship, than the puny Brat that was starv'd at Nurse. Willows are easily propagated from Willow Stakes: Lay their lower ends in Water three of four Days before you set them: let them into the Ground with an Iron Crow, but better with a Pump-auger, which loosens the Ground: A Warrener's Spade will do very well also: fasten them to a prop Stake, with wisps of Straw, and they will soon take root. H.

Page 84.6. Runcival Pease find now very little Entertainment in Gentlemens Gardens, they are however still to be seen in the Fields, as in Berkshire and Wiltshire: and are most commonly set two or three in a Hole: But in the Gardens, in their Room are got the Egg-pea, the Sugar-pea, Dutch-admirals, etc. and, with these, sticking very well agrees. A Peacock, altho' a lovely Fowl to look on, and every whit as good to eat, yet is a very ill-natur'd Bird, and particularly destructive to a Garden, as also to small Chickens, Turkey-Pouts, nay, his own kind. But seeing they are a Beauty to a House, no less ornamental than the Flowers of a Garden, and have some Skill in the Weather, etc., it may be worth while to be at some Pains to enjoy their Company, and make them less troublesome. If then you have a mind he shall not frequent your Garden, or any part of it, or any other Place, especially if it be an enclosed one: take your Opportunity, when you find him there, and with a little sharp Cur that will bark, teaze him about as long as he can stand, at least till he takes his flight, and he will come no more there: be sure to feed them well also. Turkeys, I suppose, may be served in the same manner: but the former I have known perform'd, and I have kept them with very little Dammage. H.

Page 84.7. This I take to be meant of a way of Quicksetting or fencing Enclosures out of the common Field they had in the Days of our Author: they ploughed, or drew round the Ground they intended to inclose, a very large ridge, commonly a Rod wide, and sometimes much more: this they sowed with Hips, or the Fruit of the Bramble, with Hazle-Nuts, Haws, and such like, to produce their Kind: they carefully harrowed it, and weeded it for two Years, withal ditching it well about, and in a few Years time they had a pretty Coppice, and are what we now call Shaws, and in some places Springs. This is an excellent Way to improve bleak Grounds, and it is Pity it is not continued. H.

Page 84.8. This is most in Practice in Marshy Countreys, as Lincolnshire, Cambridgeshire, and Norfolk, where the Borders of their Ditches, where the scowring is thrown out, produces Plenty of excellent Mustard-seed. It may be done in Uplands, as well especially where the Ground is in good Heart, and somewhat moist: as on the Edges of small Brooks or Drains, and will more than pay for the Labour. Where Nettles will grow, our Author observes that Hemp will grow, and kill the Nettle: he grounds his Observation (I suppose) upon the Doctrine of assimulated Juices, which the Ancients were very fond of, and perhaps not altogether without Reason: altho' too much may be attributed to it, for Nettles and Hemp are near a Kin: And I have been told by one who had experienc'd it, indifferent good Linnen may be made from Nettles, however Hemp makes better: and it somewhat reflects upon a great Part of the Farmers of this Nation, that about their Houses there are more Nettle and Dock-plots than Hemp-plots. When you sow Hemp, if your Land be rich, sow with a very plentiful Hand, your Hemp will be the finer, watch it for a Week from Pigeons. H.

Page 85.1. Vines are now to be set out: they are best propagated by slips of the last Year, with a little left to them of the Year before: we set them here in England most commonly against Walls and Houses: but if you intend to plant them as in a Vineyard, let the Ranks range from East to West. Those that thrive best with us are the small black Grape, the white Muscadine, and the Parsley Grape. Osiers are also propagated from Slips, and thrive best in the Quincunx Order: they require a Ground continually moist, and are an excellent Crop. Swans are a noble and useful Bird, their Food is the Weeds that grow at the Bottom of Ponds or Rivers: Now their Time of Laying approaches, they are naturally impatient, for though they lay nine or ten Eggs, and sometimes more, they seldom stay the hatching of above five: a Trough with Oats, placed near their Nest, may keep them to their Nests better than ordinary, for ought I know: but that, as well as the building and ordering their Nests, I leave to the more Experienc'd. H.

Page 85.2. Be sure then that your Dung be thoroughly rotten, and free from Stones: cast about now your Cow-dung and Moll-casts that lye on the Ground from your After-pasture-feed. There are many Country Fellows very dexterous at Mole catching: Some have a way of setting them with a little Dog, very neatly and diverting, to look on: perhaps, a Gentleman's or a Farmer's Time may be as well spent to follow those Fellows, while they are catching for him, as to hunt after a Pack of Dogs, or a setting Dog for Partridges, for they are dexterous at catching both Ways: and, without looking after, you may pay for Moles that never hurt you, and belong to their yearly Customers. H.

Moles, for the trapping of which each parish used to mantain a sapper and miner, are found to be excellent husbandmen, the little heaps of friable soil which they throw up, furnishing, when spread abroad, the best of top dressings. E.D.S.

Page 85.3. For killing the Mole there are several Ways, yet none, in my Opinion, come up to the common Trap, I mean the Ring-Trap, which is describ'd by Mr. Worlidge, in this manner, in his *Systema Agriculturae*, p. 216, 217. "Take a small Board, of about three Inches and a Half broad, and five Inches long: on the one Side thereof raise two small round Hoops or Arches, one at each End, like unto the end Hoops or Bails of a Carrier's Wagon, capacious enough, that a Mole may easily pass through them: in the Middle of the

Board make a Hole, so big that a Goose-Quill may pass through them: So is that Part finished.

"Then have in readiness a short Stick, about two Inches and a Half long, about the Bigness that the End thereof may just enter the Hole on the Middle of the Board: Also you must cut a Hazle, or other Stick, about a Yard or a Yard and a Half long, that being stuck into the Ground may spring up, like unto the Springs they usually set for Fowls, etc. then make a Link of Horse-hair very strong, that will easily slip, and fasten it to the End of the Stick that springs. Also have in readiness four small hooked Sticks: then go to the Furrow, or Passage of the Mole, and after you have opened it fit in the little Board, with the bended Hoops downwards, that the Mole when she passes that Way may go directly thro' the two semicircular Hoops. Before you fix the Board down, put the Hair String thro' the Hole in the Middle of the Board, and place it round, that it may answer to the two end Hoops, and with the small Stick (gently put into the Hole to stop the Knot of the Hair Spring) place it in the Earth, in the Passage: and by thrusting in the four crooked Sticks, fasten it, and cover it with Earth: and then when the Mole passeth that Way, either the one Way or the other, by displacing or removing the small Stick that hangs perpendicularly downwards, the Knot passeth through the Hole, and the Spring takes the Mole about the Neck."

Thus far Mr. Worlidge: since whom this useful Instrument has been improv'd, with some Variations: the best that I know of I had from an ingenious Farmer near Luton in Bedfordshire. Instead of the Apparatus of Board, Hoops, etc., he does all at once, only by cutting four or five Inches of Pipe, bored according to the fore-mentioned scantling of the Hoops, on one Side of which he cuts a large Notch, with a Saw directly answering to the Middle, where the Hole about the Bigness of a Goose-Quill is bored: this much better answers the Intention, than Hoops of Iron, Lead or Wood: for the Mole, once in, presses much more naturally forward, and cannot but raise the Spring, whereas they will frequently baulk the Hoop Traps. Note also, That before they are used, they, or any other, ought to be buried under Ground for some time, that they may have no exotick Scents, which is a frequent Reason why these Traps fail. As to other Ways, viz. the Fall-Trap, etc., they are not comparable to this. The burying of a live She-Mole, in a Kettle or deep Pan, I have frequently heard of, but never met with the Man that could say he had experimented it. Fuming and drowning is sometimes practic'd with Success enough. In the Year 1702, I had a Mole in my Garden, which did me a great deal of Damage, and was too cunning for all we could do: at last I found his Lodging, which was under a Stone Wall, and soon drowned him, or made him fly thence, so that he troubled me no more. H.

Page 85.4. If you have plenty of Pasture, and no better Succour for your Lambs, it is possible this Advice may not be amiss, especially to such who do not care how little Pains they take. But if you pay Money, and that a pretty deal too, for your Ground, your best way, if it be wet, is to drain it, which may be done divers Ways, according to the Fall of the Ground, etc. But suppose there is no Fall, in a dry Season dig a large Trench, like a Saw-pit, in the lowermost Part of your Ground: dig it deep, until you come to Sand, Gravel, Stone or Chalk: fill the Hole up again with Stones, over which lay Earth, and lastly your Turf: this invisible Drain will soon pay you for your Pains. This may be varyed divers Ways: however, this is Specimen sufficient for the Ingenious. H.

Page 85.5. A Lease is a name used in some Countries, for a small Piece of Ground, of two or three Acres, and certainly nothing can be of more Profit to the Farmer than small Enclosures, by whose Means he can freshen his Pasture as he pleases, his Cattle shall thrive better, eat his Ground closer, and keep their Pasture the quieter. Add to this, that by this Means he may make his coursest Meat go down, as well as the finest, and be as clean fed. H.

Page 85.6. This confirms the former, for Cattle (as all other Domestick Animals) will destroy where they have Plenty, and look sillily when they want: both which they will certainly do, if they be left to carve for themselves. The Art is therefore to deal out so to them, that their Necessity may be supply'd without Waste: and this, in this Case, is best done by small Enclosures. H.

Page 85.7. Malting is now in its Heighth, and Seed-time for Barley not far off, your Cattle call for Barley-Straw Fodder, and it is Time to think of raising Lady-Day Rent, for which nothing more proper at this Time than to thrash out your Barley: for if Barley be a Drug (as they term it when the Price is low) it will surely be so after Seed-time is over. This Rule has had indeed a grand Exception of late, but a foreign Call is not to be relyed on by the Farmer. Bread-Corn is required all the Year, and therefore very proper to be kept a little back, to employ the poor Thrasher till mowing Time comes in, etc. H.

Page 85.8. This Article is very much unregarded by Farmers at present, for fear, I suppose, of falling into Popery and Superstition: but lay that quite aside, and let us consult our Interest, Health and Gratitude. I believe most ingenious Men may easily be brought to confess, that it is to be wish'd that People would (again at this Season) refrain from Flesh, and eat Fish more frequently than they do at present: especially in those Places near the Sea, where it is very plentiful. It is our National Interest then to breed up hardy Seamen, to employ a good number of Shipwrights, and all Sorts of Handycrafts, to employ our Poor in spinning for Nets, etc., to take their Boys, when grown lusty, off their Hands, and put them to useful Employments. And it is our particular Interest to live cheap and frugal, all which may be done by encouraging the Fishing Trade, for which our Island seems adapted, better than any other Part of Europe. For it is not because the Dutch Seas are better stock'd than ours that there is more Fish brought into Amsterdam than London, London that is at least four Times bigger than it: but because there is more call for it, more Boats and Men go out to catch, more People buy it, and it is not in Power of one Sett of Men to buy it up, and throw away one half to keep the other to a Price. Every one there goes to the Market: I have seen a Burgomaster of Amsterdam go himself to the Market, when the Boats have come in, with his Silk Net in his Hand to buy Fish: And if in London People would be but at the Pains to go to Billingsgate for it, they would soon find another Sort of Provision there than there is now. And this noble Gift of God would be no longer look'd upon as a Scarcity, but a solid Support for the Poor, and a moderate Food for the Rich.

As to our Health, it is certain Flesh is more lustful and vicious at this Time than any other, and our Blood more prone to Fermentations, for which the Phlegm and Coolness of Fish is an Allay. In Gratitude the Farmer is oblig'd to eat the Fisherman's Commodity, which is Fish, because the Fisherman eats his Corn, and sometimes his Beef and Mutton. H.

Page 88 "Bleets"—the name of some pot-herb which Evelyn in *Acetaria* takes to be the "Good Henry," and remarks of it that " 'tis insipid enough". E.D.S.

Page 88. "Bloodwoort". In Lyte's Dodoens (1578) it is called also *Walwort* and *Danewort*: "The fumes of Walwort burned, driveth away Serpentes and other venemous beastes".

Page 88. "Burnet", a term formerly applied to a brown cloth, and given to the plant so called from its brown flowers. Prior, *Popular Names of Brit. Plants*.

Page 88. "Clarie". "It Maketh men dronke and causeth headache, and therefore some Brewers do boyle with their Bier in steede of Hoppes". Dodoens.

Page 88. "Coleworts". "Some write, that if one would drinke much wine for a wager, and not be drunke, but to have also a good stomacke to meate, that he should eate before the banquet raw Cabage leaves with Vinegar so much as he list, and after the banquet to eate againe foure or five raw leaves, which practice is much used in Germanie." Dioscorides (quoted in Cogan *Haven of Health*.)

Page 88. "Langdebiefe". Lyte recommends binding the root of this plant to the affected part of a sufferer of varicose veins in order to afford relief.

Page 88. "Longwort" has no particular medicinal value according to Lyte, who calls it *Sage of Jerusalem*. It was used extensively in "Meates and Salads with egges."

Page 88. "Liverwort", so called from the liver shape of the thallus, and its supposed effects in disease of the liver. E.D.S.

Page 88. "Marigolds" were considered useful in affections of the eyes, teeth and skin. Cogan *Haven of Health*.

Page 88. "Nep", or common Cat-mint, was a remedy for contusions, indigestion, pulmonary troubles and jaundice. Lyte.

Page 88. "Orach or Arach"—Recommended by Dioscorides for abdominal troubles, by Galen for affections of the throat and for jaundice, and by Gerarde for the liver.

Page 88. "Patience "—Wild Dock. Lyte remarks upon its usefulness in cases of jaundice, bites of scorpions, tooth-ache, 'King's Evil' and goitre.

Page 88. "Penny Royal" is recommended by Lyte for sweetening tainted or salt water, and by Andrew Boorde as a remedy for melancholy. It is some times called *Pudding-grass*, from its being used to make stuffings for meat, formerly called *puddings*. E.D.S.

Page 89. "Rosemary", Lyte mentions, was used for tightening loose teeth. Boorde believed in its efficacy in curing palsy, coughs and colds.

Page 89. "Tanzie". Tansy was one of the ingredients in *Tansay-Cake*, as well as a remedy for corns and a preventive from bug-bites. E.D.S.

Page 89. "Blessed Thistle", Cogan explains, is so named because of its numerous virtues, of which he gives an impressive catalogue.

Page 89. "Purslane" was a cure for erysipelas. Lyte.

Page 89. "Rokat". "Men say that who so taketh the seede of Rockat before he be beaten or whipt, shalbe so hardened that he shall easily endure the payne, according as Pliny writeth." Lyte.

Page 90. "Bassel"—Basil. Although, when mixed with vinegar and the aroma inhaled, this herb is a good restorative. Cogan quotes an unpleasant case of a certain gentleman who bred a scorpion in his brain as the result of smelling the compound too often. Hilman (*Tusser Redivivus*) refers to the plant: see page (108-4).

Page 90. "Feverfew" derived its name from having been long employed as a popular remedy in ague and other fevers. It appears to possess stimulant and tonic properties. It is nearly allied to the Camomile. E.D.S.

Page 90. "Gileflower"—Gillyflower. The spicy odour of the flower rendered it a useful substitute for the more costly clove of India for flavouring wines. E.D.S.

Page 90. "Laus tibi", "a narcissus with white flowers. It groweth plenteously in my Lorde's garden in Syon and it is called of divers White Laus tibi." Turner's Herball.

Page 90. "Nigella Romana". The *Nigella Damascena*, a favourite old-fashioned garden annual, still to be met with in gardens under the names of "Love-in-a-mist", or "Devil-in-a-bush". E.D.S.

Page 91. "Sops-in-Wine"—the Clove Gillyflower: see page 90.

Page 91. "Tuft gilleflowers". Probably some low-growing *Dianthus* such as that figured as "Matted Pinkes" by Parkinson. E.D.S.

Page 91. "Velvet flowers". According to Dr. Prior, the "Love-lies-bleeding" from its crimson velvety tassels: according to Lyte, the same as the "Flower Gentle" or "Floramor". Fr. *passevelours*. E.D.S.

Page 91. "Eiebright". "Divers Authours write that goldfinches, linnets, and some other Birds make use of this Herb for the repairing of their own and their young ones sight." Coles, " Adam in Eden", 1657.

Page 91. "Archangel". This is *Archangelica officinalis*, the stalks of which were once eaten as celery. E.D.S. Cogan advises its use as a cure for the bite of a mad dog.

Page 91. "Cummin", according to Cogan, was extensively used for washing the face, it having the effect, if not used too often, of making the complexion clear: if used to excess, it caused paleness. E.D.S.

Page 91. "Detanie". Dittany was commonly cultivated in gardens at this period. E.D.S.

Page 91. "Mandrake". They were supposed to remove sterility. There were numerous other superstitions regarding this plant: amongst others it was said to shriek when torn up. E.D.S.

Page 91. "Rew". Some suppose it to have been called "Herb of grace" on account of the many excellent properties it was held to possess, being a specific against poison, the bites of venomous creatures, *etc.*: but probably it was so called because "rue" means "repent". E.D.S.

Page 92.1. A good boyling Pea is certainly one of the profitablest Crops that belong to the Farmer, especially if they carry a good Colour: For Example, The Retailer now sells them for two Pence three Farthings the Quart, which is 2*l*. 18*s*. 8*d*. the Quarter: so that the Retailer may afford the Farmer a good Price: and it is well known they require less ploughing, less Heart, and less Inning or Harvest-charge, than Wheat or Rye, and are threshed somewhat cheaper. But a sharp black Frost will in one Night set them all going, altho' they be pretty forward: for when they are young they have the most tender and juicy Stalk of any Corn, and the Hardness of the Ground is apt to nip their Pipes in two. In Gardens they talk of watering them as soon as possible, which softens the Earth, and it is very likely may save such as are not already crush'd: but in Field-Land the best Remedy is either sowing them again, or preparing your Ground for Barley. H.

Page 92.2. It has been mentioned before, that St. Gregory is the 12th of March, Pask is Easter, Which some Years falls within a Fortnight of it: so that our Author's meaning, I suppose, is that your Mersh Grounds be not far behind your Uplands: for altho' the Winter-water lye longest upon your Mershes, yet in the Summer, by reason of their Flatness, they are more subject to Drouth than declining Grounds, and Drouth has a worse Effect upon them than on the other: they are more apt to chap their Grass, is ranker in Blade, and thinner at Bottom, than that of Uplands, and consequently more subject to wither and burn away. Fences are now much more frequent than in our Author's Time, and the Farmers more convinc'd of the Benefit of them. H.

Page 92.3. It seems, in our Author's Time, Lent was still kept up: his Book was printed in the Year 1590 [tenth edition] being the 32nd Year of Q. Elizabeth. Now from Salt Fish, Furmity, Gruel, Wigs, Milk, Parsnips, Hasty-pudding, Pancakes, and twice a Week Eggs, the Farmer's Lenten Diet, there is produced very little Dog's Meat: And a Mort Lamb now and then was very apt to whet their Appetite to Mutton, which if they once take to, there is no Remedy but hanging: Some prescribe putting him into a Stable with two lusty old Rams, who will soon give him such a Remembrance of them, that he will for ever hate the Kind: but that is to make 'em good for nothing, at best: and if you chance to suffer them too long together, that the Rams have butted themselves out of Breath, it is ten to one but you find 'em both worried. The best Way is to feed them well at home, and bury your mort Lambs in the Dunghill. H.

Page 92.4. The goeler if the yellower, which are the best setts, old Roots being red, are not near so good. Well gutted I take to mean well taken off from the old Roots: and paring is taking off all small fibrous Roots from your Sett. H.

Page 92.6. There are divers Ways of framing Hop-Hills, some are for the Chequer, others the Quincunx Form, which is, that the Hills of the second Line be against the vacant Spaces of the first: and this most follow, because the Sun has always a glade quite through. The best Way of setting them out is by a Line with Knots at the Distance you design your Hills, and pricking Sticks in the Ground where you design them, the Distances very according to the Nature of the Ground. H.

Page 93.1. Mr. Worlidge proposes six Foot at least, and in a moist, deep, or rich Mould, nine. However, the Custom of the Country, and a well-grounded Experience, are the best Guides in these cases: but be sure let not your Hills be over-poled, altho' some Hills may require twenty Poles, as well as others six or seven. Note, the Hills are no otherwise essential, than as they mark out the Place where the Hop lyes, and direct you to the manuring and poling them, and avoid the Injuries of the Foot and Spade. H.

Page 93.2. Willows are an excellent Fence for a Hop-Ground, for they break the Wind by their bending more than any other Tree, and by their wide branching they hinder the Sun the least of any Tree: they are also of quick Growth, and attract no mildews, as doth the Elm, the worst of all Trees, near a Hop-Ground. It may (for ought I know) be proper to plant some Hops in the North Fence of your Hop Ground, but by no Means on the East or South Sides. Hops will do often times very well in the Fences or Meadows and Pastures: but as I said before not under Elms, nor indeed any thing that over-shades them much, it having a strange Propensity to the Sun, and follows that Planet in its windings. In April, 1704, I poll'd some Hops, and before I had set fifty Poles, some of the Plants had clasp'd hold of their Poles and made a half turn. H.

Page 93.3. It is the Crow, not the Cross, that our Author says must be as sharp as a Stake: this Crow is to let in the Poles into the Ground, and an ordinary Crow may do without a cross Bar, if when you lift it out of the Ground you rest your Elbow on your Knee. The Hone is no other than a common Rubber, or Whetstone, to sharpen the Parer: It seems, in our Author's time, it was in fashion of the Sole of a Boot, but since there are of more commodious Shapes: the best, in my Mind, are those triangular ones used by the Fen-men and Bankers. H.

Page 93.4. For now in most Trees the Sap arises (as some call it) but more properly extends its self, and becomes more fluid. The Sap in Trees is to them as the Blood in Man, the most Sovereign Balsam for its Wounds, and is the most ready at Hand, immediately flowing to them. The Prime, as I observed before, is the first three Days after the New Moon, in which Time, or at farthest, during the first Quarter, our Author confines his Graffing: probably because the first three Days are usually attended with Rain, as has been confirm'd by undoubted Experience, whereby there is wherewithal to nourish the Plant: and also, because during the Increase of the Moon, the Vicissitudes of Heat and Cold, are not so sudden as in the Wane, the Moon succeeding the Sun after its setting, for a considerable Part of the Night: and altho' robust sound Trees may make no great Difference between the one and the other, yet these Sick and Wounded are extreamly sensible of the least outward Impression, as a Sick Person is of the Strength of small Beer, or a Gouty Person of the least shaking of a Room. That the East Wind is prejudicial to graffing happens principally from its Violence, altho' it is next to the North, the most unfertile, and brings with it very often foggy greasy Weather. But of the Winds more particularly hereafter. H.

Page 93.5. The great Question is How? For violent Winds will shake them: Birds, especially Tomtits and Bulfinches, will hang on them, and pick off the Buds: and there is no tying the Cyon, or shooting the Birds, or taking them there with Birdlime: the best Way, I know against Tempests, is to provide beforehand a Shelter against that Side from whence the Storms usually come. That impudent Bird, a Tomtit, is not easily frightened, however, if you kill one or more elsewhere, tear them in Pieces, and stick them upon Sticks near your Tree, about the Height of the Cyon, it will deter him some time, but you must expect to lose some. Beasts are more easily kept out. H.

Page 93.6. Barley is rarely sown in Clay, at present: however, some Barley Land is stiffer than other, and our Author advises to sow the stiffest first, for what Reason I cannot tell, Mr. Mortimer on the other Hand advises to sow the stiffest last, p. 107, which to me seems more agreeable: for the stiff Land may be brought to a Season, as the Farmers term it, or made finer, better when it is dry than wet. In Norfolk, near Hunston, I have seen very stiff Lands lye in vast Clods, which I was told was for Barley, and it was too late to expect much from Frosts: nothing then could moulder it but the Sun, and a very heavy Rowler. It is strange that steeping of Barley, in a very dry Season, is not more in request at present, it must certainly save abundance of Corn that for Want of it is lost. H.

"What better to skilfull" *etc.*—what can be more profitable to the experienced farmer than to know when to be bold, that is, to venture the early sowing of barley? E.D.S.

Page 93.7. The Seeds of the Weeds are in the Ground before the Barley, and the Wet brings them forward, so that they will grow faster than the Barley can. A Thistle, as far as it spreads, burns the Corn, as the Husband-man calls it, but indeed it shades it, and hinders its growth: yet these may, with Care, be weeded off with a Weed-hook, or Stabbing-knife: for they are juicy, and dye a good Way after the Wound. These are for the most Part a Sign of good

Land: but wild Oats the Peeler of the poorest Land, and who constantly attends wet Seasons, is not so easily eradicated, or any good Sign at all. They are not easily weeded when in the Blade, and by the Time they come into the Stalk they have done their Mischief. It is a wonder, not yet accounted for, how they come in such Quantities as they do in some Lands: pull one up, when in the Blade, and you will find a Seed to the Root. Mr. Atwell, in his Surveying, says he took up whole Yepsonds (that is as much as both Hands would hold at a Time) and carry'd them home: one would think they were of the Devil's own sowing, the ancient *Zizania*. May-Weed is a very stinking Weed, it commonly is brought in with Dung, but is easily weeded, and your Seed may be cleansed from Cockle with a Cockle-Sieve. H.

Page 93.8. Barly is a sprightful and tender bladed Corn, and requires as few Impediments as possible: Clods is a great one, and standing Water a worse, for no Corn is more thirsty, and bursts sooner than this: therefore the one must be broken, and the other drain'd off with Water-Furrows. There is one Annoyance I have many times wondred was not prevented more frequently than it is: The Annoyance is the Incroachment of great Roads, which in some Places increase to a vast Breadth: I know one that I believe is half a Mile broad, all in good arable Land, and consequently a great Quantity is lost in it. The common Way is to dig Trenches at a competent Distance, that the Waggons cannot go cross, and therefore the Waggons often go within them, and so more Ground is lost: Now if instead of that they would dig a Ditch and Trench all along, and fence it with Elder Sticks (which may be stuck slopewise into the Bank, two or three Foot long, making a Sort of Chequer Palisadd, and will soon grow) this may be prevented, and the Charge is not greater than the frequent digging of Trenches. In this Road, (I speak of) there is but one Hedge, of about twelve Chain long, and that has cast the Road clear on the other Side, and sav'd about three Acres of Ground, which else in likelyhood had been lost. Note, hereby your Sheep-walk is still open, and nothing will crop this Fence to hurt it.

When you have done all you can, you may safely pray God for a good Harvest, otherwise it is Mockery: and when you have it by Prayer, you will enjoy it by Praises, to him who gave it. H.

Page 94.1. After a gentle Shower, especially if there succeed a Sunshine, the Clods break best: and if the Barly be a little up it is better, rather than worse, the Horse-footing will do the less Damage: if out of the Milk, which is when the Seed grows lank, and the Root hath taken hold of the Ground, and the broken Clods refresh it. H.

Page 94.2. That is, in our Countrymen's Phrase, sow Oats, Barly and Pease above Furrow, that is upon the Land after the last ploughing, and then harrow it in: and Rye, under Furrow, that is upon the Land before the last ploughing, and so plough it in with a very shallow Furrow. Both these may, and are varyed with Success enough: but now Barly and Pease are most frequently sown under Furrow. Wheat is to abide the Winter, and if it be left a little cloddy, it will get round, and the Clods, to be sure, shelter it from Winds: it is to be reaped, so that its Roughness hinders nothing the Harvest-Work. H.

Page 94.3. If it be too cloddy, now is the proper Time to rowl it: your Rowler, for this Use, must be in the Form of a Nine-pin, small at both Ends, and bulg'd in the Middle, and then the Horse goes in the Furrow with very little or no Damage to the Corn by his Footsteps. But our Author very well observes, that unless the Weather be dry, it were better unrolled than to rowl it in the Wet. For Wheat is sown in Clay, and that in wet Weather will stick to the Rowler, and pull up more Wheat by the Roots than it will cherish. H.

Page 94.4. This our Author makes the good Woman's Care: but whose ever it is, it is at present very much neglected. It is true that the Garden ought not to rob the Field of its Time, but a little Dung can no where be better employed: And if Servants have not spare Time enough to dig it, it will pay for the hiring one to do it. A Table continually loaded with Flesh and Pudding cannot certainly be so wholsome for the Servant, or profitable for the Master, as where Flesh is allayed with Herbs and Roots, which though oftentimes at hand, Custom has brought into disuse: The Master thinks they are Sauce, and that, should he prepare them, his Men would eat ne'er the less, nay rather more Meat: and the Servant thinks nothing Meat but Flesh: So that between them, a very great Part of the Blessings of God are despis'd. I know a poor Man who liv'd near me, who was with his Family almost ready to starve, to whom one day, in Compassion to him, I told he might at any time fetch what Cabbages he thought fitting, from my Garden: his Reply was (with a scornful Smile) Cabbages, Sir, I want Meat. And indeed, the People thereabouts were extremely greedy of Flesh: eating, with great Greedyness, any thing that dy'd of it self, tho' never so Purple, and near to Corruption. The Sequel of this Fellow was (for I could name several others) that his Wife dy'd not long after, her Blood was become in a Manner wholly purulent, and vast Quantities of Matter came out of her Nose and Ears, almost incredible to relate: And I hear he is since dead, being both in the Prime of their Years. H.

Page 94.5. I have heard some say, if we had no foreign Diseases we need not any foreign Medicines: for indeed we have in this Nation abundance of excellent Aromaticks, coming little short of those we have from abroad, and perhaps better adapted to our Constitutions. H.

Page 94.6. These are, without doubt, the Situations that the Sun has most Influence upon in our Climate, and such a Declivity as the Meridian Rays are brought perpendicular to the Plane, comes very little short of the Heat under the Equinoctial: and Fertility, we know, consists in Dilatation, for which we are beholden to Heat, as Barenness, by Contraction, the Effects of Cold. Yet this Situation may not, in all Places, have the same Advantages, as where it fronts the Sea, pois'nous Mershes, Wood-bound, over-shelter'd by Woods, and the like. As also, where they lye too open, and expos'd to winds. H.

Page 94.7. Here our Author lays an Emphasis upon a Garden, which still shews, in his days, Farmers valued Gardens more than they do now. I remember Mr. Houghton, in one of his weekly Papers, advises our Farmers to put such of their Children, whom they design for Farmers, for some time to a Gardner, which would certainly give a great lift to the Ingenious, and to the dull ones no harm. Experiments may be cheaply tryed in a Garden, before they are ventur'd at in the Field: And it is no Hyperbole to say there are yet a thousand Improvements to be made in this Nation. Moreover, if my Dame be a little out of Humour, as sometimes good Dames will be, our Farmer may find no less Diversion in his Garden, than if he went a Mile or two to an Alehouse, and made the Breach wider. H.

"Without cost"—on which no expense has been incurred. E.D.S.

Page 94.8. So that your Garden brings you two Crops (besides your Bees, which may well be reckon'd a third, but of them in their Places) nay indeed as many as there are Months. For in this Month you may sow Beets, Cabbage, Carrots, Onions, Parsnips, Spinage, Garlick, Leeks, Lettice, and Pease.

In *April*, Cucumbers, Melons, Artichoaks and Madder, may be planted, and French Beans set.

In *May*, sow Sweet Marjoram, Basil, Thyme, and set Rosemary.

In *June* and *July*, Sow Turnips, latter Lettice and Purslain.

In *August*, sow Cabbage, Colliflower and Turnips.

In *September*, plant Straw-berry Setts, also Tulip-Roots.

In *October*, sow all Sorts of Fruit-Stones, Nuts, Kernels, and Seeds for Trees or Stocks.

In *November*, plant the fairest Tulips, the Weather good.

In *December*, set Beans, also sow or set Bay-berries, Laurel-berries, dropping ripe.

In *January*, make hot Beds, and sow your choicest Sallads, as Chervil, Lettice, Radish.

In *February*, sow Annice, Beans, Pease, Radish, Parsnips, Carrots, Potatoes, Onions, Parsly, Spinage, and Corn-Salading. This according to Mr. Mortimer. H.

Page 95.1. There is an Old Sawe to this purpose:

> *In Gard'ning never this Rule forget,*
> *To Sow dry, and Set wet.*

What is sown, as Seeds, are Plants compacted in a very little Space: and if they are too soon gorged with Moisture, that is faster than they can spend it upon their fibrous Root or Tendrils with which they lay hold on the Earth, they are apt to discompose their inward Parts, and, in plain English, burst. But what is sett, namely, Plants (for Beans, Pease, etc., ought not to be sett too wet, any more than other Seeds) have already Moisture in them, and their Texture is already expanded, and in its Shape: these require immediate strong Food, as being out of the Womb: and if their Nurse be dry, instead of getting from her, she sucks the little Moisture they have from them. As to the Moon, altho' I do not utterly despise the Observation of it, yet I think the best time to gather and sow is when it suits best with the Weather. H.

Page 95.2. The first Couplet has been sufficiently spoke of in the last: In the second our Author advises Regularity, which not only barely pleases the Eye, and gives an inward Joy at the first sight, but furthers the growth and Prosperity of your Vegetables. Care must be taken in this, that every Plant have its share of the Sun, of Moisture, or any other Advantage of the Ground. And such Regularity is not only confin'd to Gardens, but ought to have a Place in all other Affairs. I have heard it observ'd by some Workmen, that Turneps thrive best when houghed North and South: certainly it must be because the Meridian Sun goes more cleverly through them, and at least once a Day cherishes the Root of each Plant. H.

Page 95.3. They that sow too late have the Season following commonly too dry, so that their Seed cannot get the Strength out of the Ground. They that sow too early are as often too wet, and the Weeds grow faster than the Corn: so that here, as in most other things, both Extremes have the same Defect. Our Author's Meaning is that *Aier* and *Laier* help Practice and Wit. By Aier I understand Situation, Weather, etc.: all that depends upon the Air. By *Laier*, Composition, the Nature of the Soil, Heart of the Land, etc.: all that depends upon the Earth: These, he says, ought to be consulted with our Experience and Sense. So that what is too soon or too late at one time, may not be so in another. I know there are a great many Ingenious Men that are Farmers in this Nation, would these but set down their Observations in this kind, or communicate them, it would redound to a considerable Improvement of their native Country. Why should they bury any more their Knowledge than their Riches? Why should not the World be the better for their having been in it? H.

Page 95.4. Here I cannot but bewail again how little use the People of this Nation make of Herbs and Pulse: It is true, the Gentry use them more than ever, but the Middle sort, and Poor, think themselves undone, if they have not their fill of Wheat Bread and fat Flesh. It is not long since I have heard it spoke of, as a very ill Circumstance, that a poor Man who formerly liv'd well, hath himself, his Wife, and Children, been fain to make many a Dinner upon nothing but Burgoe, *alias*, boil'd Oatmeal: the same, to be sure, would have been said of Pease-Pudding or Pease-Porridge, without Meat, as Flesh is commonly call'd. And I do belive it is so: it is an ill Circumstance to those whose Bodies cannot bear such Food. But what Pity is it that they are not bred otherwise: How does a Scotch Man, an Irish Man, or French Man thrive in this Nation: and what miserable Wretches are our Poor, when in other Nations? how much doth the rise of Wheat or Flesh immediately affect us? of which lamentable Instances have lately happen'd. H.

Page 95.5. It is so still, and he that would think himself next to starv'd should he have warm'd Cabbage or Potatoes with his Meat at Breakfast and at Supper, shall go to Work or to Bed with his Belly brim full of Porridge and skim'd Milk: But the Error lyes in the Master more than the Servant, for other Food might be brought into request. H.

Page 95.6. The best way to destroy them is in their Nests, and then the first four are tolerable good Meat: Caddows are Jackdaws: Ravens and Rooks are protected, the one because they are supposed to eat such Ordure and Filth as would otherwise infect the Air near great Cities and Towns: the other, for I know not what. I have heard an Excuse for protecting them, I own, but it was as far off as France, and from one who I believe knew little of England: he said, that by Reason of its Moistness, England was much subject to breeding of Earth-Worms, which would soon destroy all, if this Vermine were not kept to destroy them. How judicious the Remark, I leave to the more learned Reader: I only mention it to shew Monsieur thought that there must be some Reason for the cherishing them. H.

Page 97.4. Our Author is for early Summer Fallowing, which without doubt has its Benefits: however, the Husbandman must do what is of most Importance to him, and not lose his present Barly Seed-time, which sometimes is not ended pretty forward in May. I have seen Winter-Corn, in the dry part of Cambridgeshire, very forward, which I believe was sown before Harvest: and without doubt, for cold moist Lands, it is best to be forward. Summer Fallowing not only destroys Weeds, but meliorates the Land, exposing it to the Wind and Sun, whereby it receives and is impregnated with the Nitre of the Air, as also to the Sheep, who eat up the very Roots of the Weeds: and therefore the Weeds should be turn'd up whilst juicy, or at least before they have spent any considerable Strength of the Earth. The first ploughing of a Summer Fallow, ought to be shallow, that the Sheep may come at the Roots. The Second, the full Depth, that the Air may impregnate the Mould. H.

Page 97.5. He inclines to turn in the Earth with somewhat upon it, as supposing that by the Putrefaction of Weeds, some Strength or Heart is got: but by no means it may stand until any thing run is to Seed: and some Seeds there are which are very forward. He had been himself a Farmer, and therefore very well knew, that the Farmer must considered his Circumstances beyond any other established Rule: wherefore to those, who cannot exactly follow him, he advises to do it as well as they can, and only recommends being as timely as they can, for hurrying of Cattle is by no means good, and what is got in the Ground by Exactness, may be soon lost in them. H.

Page 97.6. Of this somewhat has been spoken in former Months: neither is it impertinent here, for now the Sun begins to be somewhat strong, and that which was apt to evaporate in January, is much more now. If Fertility consists in Salts, like our Salt-Petre, as some argue, then seeing here it is in the most minute Particles, it is easily expanded by the Heat of the Sun, and the Expansion of common Salt-Petre, I am told, is above four thousand to one: so that although the Dews and Rains do bring it down again upon the Earth, it is not upon the same that it was exhal'd from, and therefore the best way is to plough it in, and secure it whilst you have it. H.

Page 97.7. Now ye may see what Medows are well laid up, and what not, and accordingly may chuse your Ground. Fen Hay, or Mersh Hay, is by no means good for a Horse, as being too frothy and light: they thrive best upon up-land Hay. A Bullock will thrive very well on Fen or Mersh Hay, and if it be Mow-burnt a little, it is not the worse, but rather the better for them, and makes them drink heartily.

Note, That this Mow-burn is such as is occasion'd by the Hay being stackt too soon, before its own Juice is thoroughly dried, and by Norfolk People is called the Red Raw: not such as is occasion'd by stacking it when wet with Rain, which is a nasty Musty, and stinks. H.

Page 98.1. The Cow, especially the common Cow, will yet gladly eat Hay: and then during the Night she can cheerfully chew the Cud, and keep her self warm, for the Nights are yet raw and cold: add to this, that where there are standing Waters (as there are in most Commons) the Cow during the Day-time licks greedily the Grass that springs through them in shallow Places, and with it abundance of Water: insomuch, as in fenny Places, they are often seen to spew clear Water. This a little Hay at Night drinks up in their Stomachs, and converts that, which otherwise chills them, into excellent Nourishment. H.

Page 98.2. The Number of Poles to each Hillock must be proportioned to their bigness, or distance from each other. I suppose in our Author's time they made the Hills less than they do now: for now 6, 8 or 10 Poles, are frequent to a Hill, some say 20 are sometimes used: However, overpoling (especially in height) is worse than underpoling. Poles ought to be set sloping, bending towards the South: and if two or three Forks be left towards the top, they prove of good use. Alder Poles peeled, I take to be the best. H.

Page 98.3. To sell to the Tanner before you are under a Necessity, is to be able to make the best Bargain: for Tanners are commonly but few in a Place, and when you are oblig'd to sell or house, may bid you a Price accordingly: However, Bark is a Commodity that at present sells very well, and Tanners are commonly pretty eager of

Buying. In Felling, he advises to cut low, for six Inches at the But, may be more worth than two Foot in another Part of the Tree. I take by breaking, is here meant sawing out, it being called breaking-up by Workmen, in those Parts near where our Author liv'd. He advises then that, in sawing-out, regard be had to cut (especially crooked Timber) to the best Advantage: as for Mill-work and Ship-work, and indeed for any Work, what is most proper for, is cutting to the best Advantage. He advises not to allow the Hewer his Chips, but reserve them for one's own Use. And here, with Submission, I take him to mean somewhat craftily: for altho' it is true, that a Hewer in some Places may make his Chips very well worth his Day's Work, yet they are seldom thrown into his Bargain, but he pays somewhat for them: Yet if a Hewer is to have the Chips at a Bargain, certain he can hew so much the squarer, and the Seller of the Timber loses all the Gain of the Wane-edges: which Gain in short is a Cheat, altho' a very customary one. H.

Page 98.4. Fencing before Felling is very proper, for Neat Cattle and Horses too, will crop the tender Sprouts of your Underwood, as it springs up, to its great Damage. Thieves have a great Advantage when they attack on all Sides, and upon a Fell they are commonly very impudent and busy. Another reason of fencing before you fell, is, that you may use your Bushes whilst they are good, which they will not be long after the Beginning of this Month: and that you may cast up your Banks whilst the Earth is moist. To stadle a Wood, is to leave at certain distances a sufficient Number of young Trees to replenish it, this is regulated by Law and Custom, only I add, that is much better to leave more than less, and that of three or four Growths, you next Fell will come by much the sooner: For as an Oak ought not to stand after he is come to his full Growth, any more than Corn after it is ripe: so methinks he should stand till then. A handsome Rank of Trees in a Hedge-row, is both comely and use-ful: and here rather than miss them, they may be indulged and made into Pollards, and they will pay well enough for their standing. H.

Page 98.5. That is the straitest, and those who are most likely and thriving, whose Root is fix'd strongly into the Ground, his But bigger than any other Part of the Tree, his Grain strait without twisting, his Bark clean without *fungi* or Toad-stools, no weeping Holes or decayed Boughs upon him. H.

Page 98.6. Elm Boards because of their large Scantling and Lightness, are commonly used for Carts, but they are very apt to warp and chop with the Sun and Weather. Ash is a tough and strait grain'd Wood, it is very apt to breed the Worms, especially if fell'd at this Time of the Year: and consequently not so fit for building Timber, as Oak, especially where it touches Lime or Mortar. But for all Sorts of Farmers Utensils, such as Plough-Beams, Axle-Trees, Spokes, Pitch-forks, it hath not its fellow: A forked Step for a Stile, I think one of the worst Uses it can be put to: for they as well as all rodded Stiles are very inconvenient, especially for the Dame and Dairy Maid. Hazel is a light Wood, and when large, tollerably strong and tough: it is much used for Forks to cock Barley or Oats, and frequently grows with three Tines, near the very Shape it is to be used in. Sallow is very light and smooth, and consequently fit for Rakes for Hay or Corn. Hulver, or Holly, is a curious fine grain'd Wood, and comes little short of Box, nay in some Respects it is better, as being much lighter and not so brittle, and yet heavy enough for Flail Swingels. Black Thorn is also very good for Flail Swingels, but more apt to spit, that is, break out in little Pieces, to these I may also add Crab Tree, which makes very good Swingels, as well as Mill Coggs, for which some account it the best Wood. H.

Page 98.7. When there is a fell of Underwood, the Buds that put out the Spring following, are exceeding juicy and tender: for had the Wood stood, they had all been put forth at Michaelmas, at the Shedding of the Leaf, and stood the Hardness of the Winter, whereby they attain a very thick Coat: but now they no sooner put forth, but they open into Leaf, and the least brush annoys them. Oxen and Cows exceedingly delight to eat them, they will refuse the Grass, to crop them, but of this has been mention made before. H.

Page 98.8. I suppose St. Foin, None-such, and several new Sorts of Grasses frequent amongst us at present, were unknown to our Author. And yet it seems by his first Verse, that in his Days they had some

Sort of artificial Fodder, perhaps Ray-Grass. The laying of Headlands for Grass is frequently used in Norfolk to this day, especially where Meadow is scarce, the like of spewy or wet Pieces among Corn: but their great Supply is Nonesuch, which takes very well in a light sandy Mould, as St. Foin in a dry chalky Soil. H.

Page 99.1. As to Commons, it is a Question whether they are of Benefit to the Poorer Sort or not? For if they are stinted, every one enjoys them according to the Land he rents, and then but little of them falls to the poor Man's Share, if not, the Rich Farmer commonly overstocks them, if good for any thing, and the poor Man has nothing but his Leavings, after he has swept it and is gone into his Ground again. And it is but very poor Milk that a common Cow gives, when she bites near the Ground: his Wife trudges Morning and Night, sometimes a Mile, and more: and if he has Children, the Eldest to be sure is kept from going to Service, or Apprentice, till they are good for nothing, and all for to fetch up this Cow, or look after the House and the younger Children, when Father is gone to work, and Mother a Milking. If they make a little Butter once a Week, he or she trudges to Market with it, and lose a Day's Work: where it is ten to one but they turn it into cheap and unwholesome Flesh. When Winter comes, he must buy his Wintering, at least with his Calf, and if his Cow come to any Mischance he is ruin'd. I am sure a very small Garden will turn to a much better Account. H.

Page 99.2. Here our Author enumerates divers Abuses of Commons as first, the encrease of a small bon'd beggarly Stock, they being poison'd with Geese, and plough'd up with Hogs, maintaining a few starv'd Ewes and Lambs, after which, as well as after the Cow, many a Day's Labour is lost, and lastly being a shrewd Means of purloining. The common Walker knows every Bodies Beast upon it, and when he sees a Stranger, he is under a dangerous Temptation, especially if it be a Sheep which may be easily carried off.

"Meet with a bootie," *etc.*—find something which was never lost. E.D.S.

Page 99.3. Our Author liv'd in the Reigns of King Henry the Eighth, King Edward the Sixth, Queen Mary and Queen Elizabeth: During which Time, there were several Commotions about the taking in of Common Field Land, which I find our Author entirely for, as being for the undoubted Interest of the Nation: for in short, the greatest Part of the Privileges of common Fields, *etc.*, are but so many Privileges to wrong and quarrel with their Neighbours, to foster a litigious Humour, and set them together by the Ears: to breed up a starv'd beggarly Stock in Hopes of a Fortnight's Food, of which before. The continual Work that Enclosure produces, is certainly of more Value to them, and the Haws, Acorns, Crabs, and Mast of a Hedgerow, will twice countervail the Shack of a Field: Besides, if the Hog be kept out the longer, the Gleaner is not, which turns to most Advantage. H.

Page 99.4. And yet it is but in very few Places, that they will agree to have a Swine-Herd, some for fear of its being the Occasion of a Stint, or setling at the Court what Number of Hogs each shall keep: others in plain down right Terms, least they should not trespass. I know one who us'd to brag she had the prettyest Creatures (meaning her Swine) who would lie out sometimes a Week together, but then came home so fat and in so good likeing, it did her Heart good to see them.

This is what must exasperate any Gentleman, or Farmer, to Sallow or Shake them soundly with his Dog, and not value the Noise that either of them make, for it is an extravagant Damage that a Hog will do in a little Time, especially amongst Sheaves: The poor Man pays for this too, he must have Pease to fat them after all this: and either the best Part of his Harvest-Money, or his Winter's threshing must go, and if he sells, the Butcher will give him little Profit. Yet I am not against a labouring Man's keeping a Cow and a Sow, provided the Milk be used in his Family, and his Pigs sold for Roasters, and that he rather buy Shots, (Pigs about four Months Old) than rear. H.

Page 99.5. Folding of Land is one of the most ancient and ready Ways of dunging: and will serve very well for two Crops, but it cannot be had by every one, especially Sub-tenants, who live under a Landlord, or Farmer, who keep a Flock: they will be sure to fold their own, and rarely will be hired. However, if they feed upon the Ground, they commonly leave the Price of their Food behind them, and that is some Benefit, provided the Shepherd keep them together, and make them go regularly over each Ground, but it is too often otherwise, now as well as then: and if the Farmer do not mind his Shepherd, he will as often trespass upon his Master as any body else. H.

Page 99.6. This at first Sight seems somewhat conceited, but considering the Ease wherewith such a Thing may be done, the meaning is good. What if the Plough-Boy pick a Wallet full whilst the Plough-Man is untracing the Horses? What if the Shepherd, who spends half his time in Idleness, employ some of it in picking Stones into Heaps, where they may lie until the Cart is at leasure to fetch them: this is as easy, and as much in Sight of his Charge, as in Nut time to fill his Pockets with Nuts? Now where Stones do annoy the Land, and it is found worth while to employ People at Wages to pick them off, certainly it is worth while to pick them and bring them home at spare Times, for let them be never so troublesome abroad, there are Uses enough for them at home. H.

Page 99.7. Suffolk and Essex were the Countries wherein our Author was a Farmer, and no where are better Dairys for Butter, and neater Housewives than there: (if too many of them at present do not smoke Tobacco). Their Butter has a Smell and Flavour beyond any thing to be met with elsewhere; and by August it shall acquire a Firmness or Hardness, and be fit for potting. I can assign no better Reason for this, than the Number of Cows they keep, and the smallness of their Inclosures, by which Means they have frequently fresh Pastures: for when a Cow bites near the Ground, she neither gives in Quantity or Quality her Milk. I cannot deny, but there may be something in their Breed, and I know that one Cow will give much better Milk than another, altho' in the same Pasture, the Champ or Feed may also contribute much. Rampions, Saxifrage, and no doubt many other Grasses, as St. Foin, etc., give an odd unfashionable Taste to Butter and Cheese, and by consequence there are those Grasses which please our Palates as well. H.

Page 99.8. The Eye of the Master makes the Horse fat, and that of the Mistress keeps her House and Dairy clean: without due care and following Servants ever were, and ever will be, lazy and liquorish. Cleanliness and Opportunity are the two Supports of a Dairy, and if it is the Servants Business to act, it is the Mistresses to contrive. H.

Page 100.1. So far from Gain, that he that trusts to unfaithful Servants, shall certainly be a loser: it is incredible the Waste that they will make, where left to themselves: I know an Estate now worth 200 *l. per An.* and when in Servants Hands never made so: nay, was sometimes in debt, and the worst is, the Fault is remediless: for if the Dairy-Maid Cisley, or Plough-Roger, do go off with somewhat more than their own, all the Redress is, being at more Charge, at last perhaps they are whipt, which is your utmost Satisfaction. H.

Page 100.2. "Waine her to mee:" Perhaps—waggon, that is, drive, carry her to me. E.D.S.

But notwithstanding the Greatness of the Provocation, if a Servant be punished, perhaps you may stay long enough for another. Wherefore, a Master and Mistress's Diligence and Instruction, is more than doubly rewarded. An indifferent Servant shall mend under a diligent Master or Mistress: but under a slothful and careless one, the best is sure to be bad. H.

Page 101.1. Floting is taking off the Cream: some, as in Devonshire, scald their Milk before they flote it, and this raises indeed the more and thicker Cream: but the remainder to be sure must make miserable Cheese: In Suffolk they are also noted for this fault. In Holland they have an ingenious way of making their Skim-Milk-Cheese eat tollerable, namely, by mixing it up with Seeds, and this scrap'd and eaten with other Cheese, gives a Relish good enough. H.

Page 101.2. Formerly when Salt was cheap, some salted with a plentiful hand out of Covetousness. H.

Page 101.3. Because she did not work the Curd well together. H.

Page 101.4. The Curd was not well broken. H.

Page 101.5. Toughness is occasion'd by its being set too hot, or not wrought up, and the Curd broke in good time. H.

Page 101.6. What he calls Lazer, which is an inner Corruption, or Rottenness of divers Colours, is chiefly occasion'd from their using

Beastings, or Milk soon after Calving; which altho' to it, as well as Butter, it gives a very bright Yellow at first, soon corrupts and is unwholesome. The blew Mould is occasion'd from Moisture, and Cheeses touching one another, the brittle Mould from Bruises, the Cheese-cloths being not clean, or sower, gives a bad Taste also. H.

Page 101.7. A Slut indeed, but Wenches when they can get a Looking Glass, will be running into Places where they are least suspected, and be combing and tricking themselves up: and therefore it is not without reason, some neat Housewives cannot endure a Looking Glass to hang over a Dresser. H.

Page 101.8. If the Curd be very well wrought before it is put into the Press, it will need much the less. Some there are who lay no Weight at all upon them in the Press, but work them very well before hand. H.

Mary Magdalen is introduced to express curds ill separated from the whey. M.

Page 101.9. Gentils comes from their being kept too moist and warm, too seldom turn'd, and too soon heap'd one upon another, and perhaps from being Fly-blown. H.

Cheese full of gentils is deemed to be fit only for the magpie. M.

Page 101.10. When the Bishop pass'd by (in former times) every one ran out to partake of his Blessing, which he plentifully bestow'd as he went along: and those who left their Milk upon the Fire, might find it burnt to the Pan when they came back, and perhaps ban or curse the Bishop as the occasion of it, as much or more than he had bless'd them: Hence it is likely it grew into a Custom to Curse the Bishop when any such Disaster happen'd, for which our Author would have the Mistress bless, *Anglice*, correct her Servant both for her Negligence and Unmannerliness. And indeed throughout this Author, it appears that Farmers, like Masters and Dames, might, and did correct their Servants, and were not oblig'd to treat those like Gentlefolks, who could not be suppos'd to have any Civility or good Breeding. H.

Page 103.9. "Take nothing to halves"—do nothing by halves. E.D.S.

Page 103.12. "Tell faggot and billet" *etc*.—count your faggots and firewood to prevent the boys and girls from pilfering it. E.D.S.

Page 103.13. When the coals are delivered see the sacks opened for fear the coal dealer and the carman should be "two in a pack", or "harp on one string", and between them you be defrauded. E.D.S.

Page 104.1. When flocks were more uniform as to breed and management, lambs used to be separated from their dams about St. Philip and St. James' day, May 1, for the purpose of tithing as well as milking. In regard to tithing, no specific day can now be fixed: and as for milking ewes after the lambs are weaned, the practice has become almost obsolete. While it existed, no doubt it was prudent to warn against the consequences of continuing it too late in the season, lest the ewe should thereby be weakened and rendered incapable of supporting the severity of the winter. M.

Milking of Ewes is now very little used in the Southern Parts of England, and not so much in the Northern as formerly, it being of all Milk accounted the worst: and, by reason the Ewes must be milk'd backward, the uncleanliest. However, if you intend to sell your Lambs off at some of the May Fairs, it is time to teach them to feed themselves. As to leaving off milking at Lammas: I take it, there is no necessity of being precise, for they grow dry of themselves very soon after they have taken Ram: and I take it, there is no Danger at all, or fear of singing *Requiem Æternam*, as our Author terms it, if they be milk'd, (or which is the same) if their Lambs go by their sides until that time, or some time after: for sucking certainly keeps them from the Rott: And there is nothing more dangerous to the Ewe, than to grow fat soon after taking Ram, or to be in plentiful Pasture until about a Fortnight before yeaning. Of the time of their taking Ram I suppose we shall more particularly speak hereafter: I shall only therefore here insert this general Rule, namely, That the best time for Ewes to yean in is when the Farmer hath plenty of Food and Succour for them (however, the earlier the better) and by consequence the best time for them to take Ram in, is just Twenty Weeks before that time. H.

Page 104.2. Folding and Milking at the same time is, without doubt, too much: for altho' folding is very beneficial to Land, there is none but must own it is prejudicial to Sheep, especially on moist Lands, and in wet Weather. However, if Sheep be well fed, or (as our Author terms it) have Pasture to fulfil their Desire, they may bear what Hardship you put upon them the better: But such Pasture consists not only in Quantity but Quality. Your Sheep every Morning come hungry out of the Fold, and fall greedily upon what they first light upon, which if there be no farther Care taken, may be as well bad as good: whereas they ought to be drove immediately to the sweetest and dryest Champ, such as Broom-Furze or Juniper. H.

Page 104.3. Our Author, I suppose, took this for a considerable Secret: for if Ewes Milk be fit for any thing it is for Cheese, of which I have eaten very good in Dantzick: And without doubt a skilful Hand may so qualify it with Cows Milk as to take off so much of its rankness as may bring it to a grateful taste. Some will have it that Parmesan Cheese is a mixture. H.

Page 104.4. If Sheep or Lambs are at any time laxative (which they will be whenever their Food is too moist) then their Dung hangs to the Wooll, and there breeds a Worm which soon seizeth the poor Creature in his Rump, which is a very tender part: and this without doubt makes him uneasy, which he shews by the wrigling of his Tail: These Lumps or Treddles being barberly cut off, that is very close, and the part rub'd with Dust, was in our Author's time the Cure: The common way now is, after the Treddles are cut away to anoint them with Tar: or, if the Maggots are got deep into the Flesh, to wash them well with Scab water, namely, a strong Decoction of Tobacco-Stalks in Chamber-lye. H.

Page 104.5. Reeding is no where so well done as in Norfolk and Suffolk, and is certainly, of all covering, the neatest, lightest, and warmest: neither will it (like Straw) harbour any Vermine, and besides comes very reasonable and cheap. If it be now and then cleansed from Moss, which stops the Water and rots it, and smooth beaten, to be sure it will last the longer: but it is not very apt to gather Moss, and will bear a better Slope than any other Thatch. H.

Page 104.6. By Woodsere is meant decay'd or hollow Pollards, which he advises by no means to lop at this time, for it is indeed the ready way to kill them, or any Tree, altho' pretty sound. Ivy sucks not only from its Roots, but by adhaesion having as many Roots as Tendrils, by which it cleaves to the Tree, and hinders its addling, *Anglice*, being added unto or increasing in bulk. H.

Page 105.1. Threshing of Corn hath for a long time been, and still continues to be, the way of cleansing it from the Straw and Chaff: and altho' many other ingenious ways may be found out to perform the same thing, I am apt to believe there is none but will be attended with more Inconveniences than this, especially as our Farmers Circumstances now stand: for the Thrasher doth not only thrash, but serves the Cattle with fresh Straw, the Hogs with Risk (Offal-Corn and Weeds, and short knotty Straw) the Poultry with Seeds and Pickings, who all constantly attend on him, are under his Eye, and he is always at hand, ready upon any Emergency of Fire, Thieves, sick Cattle, *etc*. H.

Page 105.2. Our Author means the Winter is not yet gone, and therefore some dry Meat must still be kept. The Nights are yet sharp, and tender Cattle must be housed. Land-Floods are very apt, about this time, to overflow low Grounds: And in most Uplands there is very little Bite. H.

Page 105.3. We sow now much earlier than we did in our Author's time, so that our Wheat in May is generally too forward to be eaten down: and as for mowing it, I believe it is very little practis'd. This is certain however, that where the Ground is too rank or lusty, neither is the Corn so good, for it runs more to Straw than it should: and it is very subject to be irrecoverably lodg'd: Irrecoverably, I say, because shorter Straw may rise when the Corn is much forwarder than longer Straw: and if it should not lodge, but be ripe sooner than the rest of the Field, the Birds to be sure will be first there. H.

Page 105.4. A Weed-hook is an Instrument well known, and therefore needs no Description, but a Crotch I take to be an Instrument of this shape, put to a handle of 4 or 5 Foot long, now not much used, but for ought I know may find Acceptance with some, and therefore have here inserted the Shape. There are many other Instruments for weeding, according to what Weeds they are to extirpate, and the Ingenuity of the Farmer. I knew one who had a Field of Wheat overrun with Cleavers to a prodigious Degree: the Wheat was near earing, and the Cleavers clang so to it, and ramp'd so high, that it was impossible, if they had gone on, but the whole Field must have been an entire Matt: The Farmer set his Wits to work, and made a sort of a Rake, but with Teeth about two Foot long, and the Handle not much longer: with this he comb'd his Wheat, as one would comb a Head of Hair, from the Roots upwards, and by this means destroy'd the Cleavers, and had a very good Crop. H.

Weeding of winter corn certainly should not be delayed longer than this month, though the lenten crops may be cleared during June. M.

Page 105.5. The Farmer has a great many Enemies, and of them Weeds are none of the least, particularly these here mentioned: as, The May-weed, which is almost to look at like a Camomile, but a filthy stinking Weed, and burns, that is, spreads itself to such a Compass, as kills all the Corn near it: it is frequent where old Dunghills have stood long, and consequently lives upon the best, and sucks the very Heart. The Thistle is also a Sign of a good Soil, but is a very bad Guest, and must be destroyed in time, for if he be suffer'd to seed, the Seeds flye and infect the Summer Fallows. The Fitch, or as some call it, the Tine-tare, is common upon almost any Land: he spares not the poorest and hungryest, and must be weeded in time or he pulls down the Corn. The Fern, or Brake, is a very bad Weed where it takes, and not easily weeded out: it is observed they dye pretty far below any Bruise, and therefore some advise to mow them down, when they are yet young, with a wooden Scythe. The Cockle has for a long Time, lain under a bad Name, but, to give him his due, he is not so pernicious as these his Companions: 'tis true, he (as all other Weeds) will live upon the best that the poor Ground has, but he spreads not much, is easily weeded by hand, and his Seed easily separated from the Corn by the help of a Sieve: Nay, grind him he gives a White Flower, malt him he works with the Barley: however, his Room is better than his Company. Boddle is a Weed, like the May-Weed, but bears a large yellow Flower, and is a very filthy spreading Ulcer upon Land: it is hardy, and will grow again, unless the Roots are clean pulled up: the Seed is also very spreading. H.

Page 105.6. The Weeds, if neglected, rob the Corn both in Quantity and Quality, increase the Husbandman's Labour, make him run greater Hazards than needful (for he cannot inn weedy Corn as he can clean) and run down his Market: this is in Proportion as 1 to 32, if not more. What is intended for Seed to be sure ought to be thorough clean. H.

Page 105.7. This useful Grain is very much disused in England, I suppose because of its Rankness of Taste, which in my Mind is not unpleasant. It is for the most Part given to Hogs and Poultrey, where it has no good Reputation, for it makes the Fat frothy and light, and apt to drip away. But then methinks it should be the better Food for Man, to whom too much hard Fat can be no Benefit, but a Burthen. Excellent Pancakes are made of it in Holland, and are eaten by the Best: and perhaps other Wheat had never rose to so great a Price (as it did here of late) if People would have made shift with any thing else. It will grow upon dry and poor Land, but must be sowed late, because it cannot endure the least Frost. It is frequently ploughed in, when in Blossom, to make a Season for Wheat the ensuing Year. H.

Page 105.8. It is also very proper to sow it before Wheat, the Ground is made clean and fine by it, and it sufficing itself with a Froth leaves the solid Strength for the Wheat. H.

Page 106.1. This Observation I take to be of very little Use: for the latter End of May is most commonly dry, and very unfit for sowing Pease, which require a moist Earth. Pidgeons, Rooks, and other Vermine, about that Time begin to be scanted, and will certainly find them out, be they in never so by a Corner. If they are much shaded (as by the Word Corner I suppose he means) they will run to nothing but Hawm. And lastly, if they do come to Perfection, and are fit to eat in Harvest, the gathering and shelling them is more worth than twice their Value. I suppose, in our Author's Time, French or Kidney-Beans were not so well known as they now are. H.

Page 106.2. I have spoke elsewhere somewhat on this Subject, and therefore shall only observe here, that it is a great Pity that so much Money goes into foreign Parts for that, which with Industry, we might as well have at home: we have Ground every whit as fit for it as any where, and People as ingenious, and Winter-Evenings Work as much wanted. The Fimble, or Female Hemp, is that which is ripe soonest, and fittest for spinning, and is not worth above half as much as the Carle with its Seed. H.

It is curious that the Karle or male hemp should be in reality the female plant, but other authors use the names in the same way. E.D.S. (See note to 116.2).

Page 106.3. The Hop-Yard must now be minded, and the Hop guided to his Pole, those who are unruly must be bound with Woolen Yarn, Hemp, Peelings, or Bast. I am inform'd that twenty Shillings an Acre is the common Price for looking after a Hop Ground. H.

The fact of the hop being one of the plants which twine from left to right had thus been observed as early as Tusser's time. E.D.S.

Page 106.4. Here he enumerates some of the poor Hop's Enemies, at least such as may be remedied, which the Weeds may be by paring the Ground if the Season be wet, or if dry by houghing it. How the Peacock may be frighted from any Place I have mentioned before, and I suppose the same Remedy will serve for the Turkey: I have experienc'd, they are very great Enemies to the Hop at this Season. H.

Page 106.5. This Custom of taking of Wheat to get out the Tine-Tare is very little practis'd at present, neither is it very proper, unless a Ground be in a manner quite over-run with it. The better Way, I take to be what he orders for Rye, which he supposes too forward, to rake, namely, to break the Tine off at the Root, and to let it stand on the Straw: for it sticks so close, and is wound so often about the Straw, that it will be apt to tear the Corn up by the Roots rather than come off. H.

Page 106.6. If the Quickset be laid in the Bank, it may most easily be done by a Boy going along the Ditch: in it is true, after a Shower the

Weeds come up best by the Roots: but never stay for that, a Boy that will work may easily weed forty Rod in a Day. H.

Page 106.7. For if the lower Drains are not kept open and free, the back Water is kept longer than ordinary upon the Upper-Grounds; it's true, if it is kept too long, it does loosen and soften the Sward, makes it subject to Rushes, Arsmart, and coarse Grass. But latter Experience has taught us, that at this Time of the Year such Ground as is intended or laid up for Hay, will endure (nay requires) a pretty deal of Moisture, and a Stoppage below, wisely manag'd, may be of as good Use as draining. H.

Page 106.8. The Proverb says, *A Swarm in May is worth a Load of Hay*, so that our Author speaks modestly when he values them but at a Crown. Their Hours of swarming are for the most Part between the Hours of ten and three, and ought to be watch'd every Day: which may be done by a Boy or Girl, that at the same time may spin, knit, or sow. The tinkling after them with a Warming-Pan, Frying-Pan, or Kettle, is of good Use to let the Neighbours know you have a Swarm in the Air, which you claim where ever it lights, but I believe of very little Purpose to the reclaiming the Bees, who are believ'd to delight in no Noise but their own. H.

Page 107.1. In stiff Ground, if a dry Time comes, though your Plough and Team may be very good, yet the one may be too slender, and the other too weak: and if this happen in the latter End of May, 'tis ten to one but it lasts a good Part of June. All this while your Ground is spending itself in Weeds, and you lose the most proper Time to kill them if your Ground had been turn'd up. H.

Page 107.2. Concerning dunging hath been disserted before: and I believe the last Line of this Stanza should read, *More profit the sooner to fallow* (not *follow*) *thereon;* that is, the sooner you plough it in the better. H.

Page 107.3. That is, if you have Muck to spare make your Dunghil upon a Head-land, it is nearest the Gate perhaps, and is dripped and shaded: so that the Strength will not exhale, but rather increase by the Addition of Moisture. H.

Page 107.4. Without doubt, the best Time for picking of Stones is when the Ground is Summer-fallowed, especially after the Second ploughing, which turns up deepest. About this Time also Highways are mended, and Stones are wanted. But his first Line, altho' perhaps only made for Rhime sake, is what I take most notice of: I would fain have Children hired and encouraged, as much as possible, to lay to their Bones, and be able betimes to do and endure. The poor Man complains of his hard Fate, envies those who live easier than himself (as he thinks) and resolves his Son shall not be such a Slave: Whatever it cost him, he will give him Learning. He does so, and makes this Creature, that might have been as useful as himself, an idle, malapert, conceited Wretch, that thinks himself learned, because he can read and write and his Father can do neither: whom he scorns and despises for his Cost and Care, and thinks labour beneath him. These are the Pests of all well-order'd Governments, and those who furnish Prisons and the Gallows. It were to be wish'd that every one had a competent Stock of Learning (Reading and Writing, I mean) it would make the Thing more common, all Men more useful, and take off that false Value some put upon themselves. And it is as much to be wish'd, that with that Reading and Writing something solid might be taught, some mechanical Employment that might employ that Reading and Writing: at least, give the Child a Taste of the Use for which his Learning is intended. H.

Page 107.5. If the Mother and they are within hearing of one another there will be nothing but perpetual Bellowing and Din, and neither of them will take their Food contentedly. H.

Page 107.6. Nothing that is young ought to be pinch'd of sufficient Food and Sleep, and therefore in your Barth there should be always clean Water standing by them, for they will frequently get up, drink, and lye down again. In frosty Weather it is not amiss to break the Ice for them every Morning: for they are a silly Creature, and when they go to drink, and find the Water dry, they are apt to refuse it some time after. And that there are frequent Frosts in April and May, any one who gets up betimes may be convinc'd of. H.

Page 107.7. Whoever denies his Beast Plenty when God sends

Plenty, must expect he will not be able to endure Want. The forward Summer Food is what fills the Veins with Blood, and consequently covers the Body with Fat, which is not only a Covering, cherishing vital Heat, and defending it from the Injuries of the Air, but it is a Store, a Store of Food against ensuing Scarceness: Whatsoever poor Beast is depriv'd of these, his Winter Food and Clothing, must be in a wretched Condition: when he must struggle with Scarcity and Cold: his coarse Food will then want Heat to digest it, and he shall starve upon what plumper Cattle will thrive upon, and the Churl his Master deserves to lift at his Tail, or worse. H.

Page 107.8. The Fewel here meant is such Wood as hath either been felled or grubbed during the Winter, which is well know never to get by laying abroad. H.

Page 108.1. In our Author's Time, and not long since, the Yarmouth and Ipswich Colliers were laid up in the Winter, and then the Spring-Market was always dearest and the Summer cheapest, but since, that Affair is very much varied: however, Carriage is best and cheapest in Summer still. H.

Page 108.2. This alludes to the Custom of Norfolk, where the Dame and her Maidens get up very early to their Dairy, on churning Days, and are as duly laid (as they call it) sometimes from eleven till two. The Ploughman takes two Turns, or Bouts, the first from betimes in the Morning until about eleven, and after his Dinner and Nap (which sometimes lasts till two also) he takes a fresh Pair of Horses and ploughs until Night. How good a Way this is I leave to those who have experienc'd: It looks indeed lazy, but, to give them their due, they are an active People enough: for at mid-August, or their Harvest Time, one would think they never slept at all, there is of them all Day long in the Field, and during all the Moon-shine of the Nights. H.

Page 108.3. Roses, Mints, Balm, and some other Aromatick Herbs, give very pleasant and delightsome Waters, if skilfully drawn off: but the numerous Catalogue of simple distill'd Waters, especially if drawn from the cold Still, are for the most Part somewhat worse than fair clean Water, and will corrupt sooner. Our Farmer may, with a good Alembick, distil the Lees of his strong Drink, Metheaglin, and Cyder: and if he has too many Goosberries, with a very little Trouble he may get a good Spirit from them also: and when he has done, the same Lees and Goosberries, *etc.* are rather better for his Hogs than they were before. Such Spirit he may again rectify over Wormwood, or what else he thinks fitting, and then he has a good Dram at Hand. H.

Page 108.4. This, I suppose, is a Complement to the Farmer's Landlady, or any other Lady that visits his Farm: for most People stroak Garden-Basil, which leaves a grateful Smell on the Hand: and he will have it, that such stroaking from a fair Lady preserves the Life of the Basil. H.

Page 108.5. To profit is a modest Word for to Bull, and the Scope is, he would not have the Farmer suffer his Cow to be tantaliz'd with an Ox, for Oxen are somewhat gamesome at this Time of the Year: tho', by the by, 'tis inserted here somewhat *mal a propos*. H.

Page 110.6. Running Water to be sure is best, for it is a vast deal of Filth that washes off from a Sheep: but then it is not oft times very sheer, and cold, especially in small swift Brooks. After Washing, some good swarded Pasture is best for them, provided it be fresh and not too near the Ground. Keep them as much from Paths and frequented Roads as possible; for altho' some pretend that the Sand makes the Wooll weigh, it is a Cheat, and makes it shear the worse, and what is got that way, may soon be lost in the Life of the Sheep: for the Workman finding double the Trouble, will soon grow careless of their Hides: besides the Price of the Wooll, that being run down in the Market. H.

Page 106.7. A Slash is bad, but if well covered with Tar in due Time, is soon cured: but a Prick with the Point of the Sheers passes oft undiscover'd, which swells, putrifies, and oft-times destroys the poor Creature. H.

Page 106.8. This is to be understood of the second Year after they are yean'd, for then they are yet much tenderer than the other Sheep, and therefore to be shear'd last: for if they are shorn whilst the Nights are cold, they will be apt to be stiff, and not able to rise in the

Morning, when Mr. Magpye will be sure to be with them betimes, and pick out their Eyes before they are stirring. On the other Hand, to leave the Wooll on too long, is to trouble the Creature with an unnecessary Burthen to hinder it from cleverly stooping to its Meat, as well as walking about to seek it, and to mat the Wooll so as to be good for little. Every Thing has its Time for Ripeness: and when ripe, it ought to be gather'd in the best Opportunity. H.

The E.D.S. note on the magpie allusion does not agree with the above explanation: "that is, the magpie will save you the trouble, etc., alluding to birds eating vermin on sheep's backs."

Page 111.1. Where Land is likely to burn, such as hanging Sides of Hills, gravelly Ground, and the like, if the Weather hold dry, mow it ere it begin to wither. Lower Grounds may go longer, but then not only (as our Author advises) cock against Rain, but in the fairest Weather, towards the Evening, and that before the Dew falls, whilst the Heat of the Sun is yet in it: and in so doing, your Hay shall make during the Night as well as the Day. If Hay be hous'd or reek'd too green, provided it has not taken Wet by Rain, it is apt to Mow-burn, and sometimes sets it self on Fire, which shews it is at Work all the while: thereas Hay made up wet by Rain, shall turn to a filthy stinking Mould. Note here, although Mow-burnt is an extreme, yet there may be some Degrees of it very useful, particularly if your Hay be coarse, Mow-burning it a little tenders and sweetens it. I have known near the North Bank, between Wisbich and Peterborough, good Hay for Cattle made of mere Sedges, after this Manner. II.

Page 111.2. The Grass and Ground ought to be very dry, before you begin to make Hay. Till which Time, you may employ your Team and Servants in Summer-fallowing, carrying Muck and other husbandly Matters: So that you may set forward your Affairs in such a Manner, that when Harvest time comes, you will have nothing to do but to tend it.

Your Horses are now also in very good Case, and if you have not Work for them at home, a Bargain of Timber-Carriage is not amiss at this Time of the Year: or any other Work that brings Money into the Pocket. H.

Page 111.3. He that goes a Borrowing, goes a Sorrowing: however, there are some odd Things that it is hardy worth while to provide ones Self with (and where others who have more Occasion for them are willing to lend, such as Ladders of extraordinary Size, Draining-Ploughs, Rook-Nets, *etc.*) they may be dispensed with. But what is for every ones Use at the same time as Rakes, Pitch-Forks, Syths, Carts, Waggons, *etc.* it is unpardonable in the Farmer to be unprovided with them, and the Lender's Answer ought to be, *I want them myself.* Moreover, as our Author well observes, besides the Payment, the Courtesy will be required doubly: and who would willingly for every small Matter be under such an Obligation? Who, but such as are wilfully lazy? And they are those who indeed take most Pains. H.

Page 111.4. It is too late to be Mending, when the Cart should be a Working: in Hay Season you ought (if possible) to be too quick for the Weather: at best your Time of Carriage is but a Part of the Day, for Mornings and Evenings are unfit, and that Part of the Day that is often catching: So that altho' the idle Carter swore his stinking Breath away at you Importunity, it is not amiss to follow him, and see that all his Tackle be in order. In Corn Harvest, the Clefts at the Bottom of a Cart or Waggon, may give the Goose or Hog more when they have enough: but a close Cart will save more than the Flesh of one Hog or ten Geese are worth. H.

Page 111.5. The Sun does more Harm to a Cart than either Wind or Rain: however, they are all three Enemies, and are easily prevented by a Cart-shed, which need not cost much, for one may be made with eight Crochets (forked Posts) and as many Spars: It may be covered with Bavin Wood, Brakes, Furzes, or other Firing. However, a handsome Cart-house, with a Granary over it, is better: Under these a Cart is immediately out of Wind and Weather. Your Hog (a Creature extreamly fearful of Wind and Rain, and to whom the Heat of the Sun is very pernicious) finds here immediate Shelter and Shade, and a Wheel to rub against. H.

Page 111.6. In the Margent our Author explains a Hovel to be a Place enclos'd with Crochets and covered with Poles and Straw:

These are of very good Use to put Corn-Stacks, especially Pease and Tares upon: for if there be a Dog Kennel under them, they are hollow under, free from the Damp of the Earth, which they are very apt to draw, and out of the Hog's Reach, who will certainly undermine them, if he can. H.

Page 111.7. The Use of Barns is in most Request in the Southern Parts of *England;* and altho' they are very useful and convenient for the Tenant, they are very chargeable to the Landlord: for this is certain, the more Building the more to be built, or at least to be kept in Repair. But Landlords are for improving their Estates (as they call it) that is for great Rents, though they purchase them: for when a Thing is to be hard let, a Tenant is in the Right to insist upon his utmost Conveniencies. Now supposing a Tenant has a good Bargain, and is loath to be craving, I assure him very good Shift (in a considerable Farm) may be made with a small Barn-Room: and Reeks and Hovels have their Conveniencies, as Corn doth not Mowburn so soon in them as in the Barn. Hovels may be made so as to afford no Shelter for Rats and Mice: and by the Help of an old Sail to clap over them till they are compleated, your Corn may be as free from the Accidents of Weather, as in a Barn: only take Heed, if you thatch them, that you watch the Thatcher that he wet not his Straw, for if you don't, he certainly will, and that will musty your Corn a pretty Way. Wherefore, some, with very good Reason, never thatch their Hay-Stacks, but make them with a very keen Slope, and rake them well down. H.

Page 111.8. That is, lay it in the best Place you have, for which the Wheat-Houses now in request (and which are much easier seen than described) are I think the best. Mustiness in Bread-Corn is not to be endured, and wherever there is the least Drop of Moisture, it must be expected: Neither is it very excusable in Pease and Fitches, for a Hog is as nice when he comes to be fatted, as he is greedy when he is kept hungry. H.

Page 112.1. Whins and Furzes are the same, and the Sides of a Hovel wattled with them will keep out a pretty deal of Weather: but I take not that here our Author's Meaning, but that on each Side and on the Top of your Hovel, a Stack of Whins, Brakes, or whatever other light Firing you have, be erected. This, as you consume (being very good for Baking and Brewing) renew again, because he would have your Turf and Seacole, tall Wood, or Bavin and Billet secured under, or indeed any thing else: as for Example Reed for Thatching, which altho' perhaps bought in only for Rhime sake, may be here secured from the Weather: a very few Crochets and Poles will make up such a Hovel, and those very slender ones too. Besides these, your Yard may be fenc'd in with this light Firing, a Yard or two thick: and this in Bleak Situations, as Warren-Houses, *etc.* is an excellent Relief for Cattle in Cold Nights: So that with a very little Pains, nay none in Comparison to Ricking, the Husbandman and his Cattle are warmed with the same Firing. H.

Page 112.2. It is very needful for a Farmer to have some smattering of the ordinary Trades, and not send to the Carpenter and Collar-Maker, or run to the Smith at every Turn: Their Time is oftentimes more worth than the Job, and Goings and Comings must and ought to be reckon'd for. Besides, sometimes a small Job to your Plow, or Cart, a Stitch or two in your Harness, or a Nail or two in a Horse's Shoe is required in an Instant, when your whole Team lose their Time too, whilst you send abroad. A Stitch in Time saves Nine: and the Woman shall look much tighter who herself takes Care she be so, than she that trusts to any other to keep her so. I have known in a very inconsiderable Farm, the Bleeding of Horses come to a Sum: and all this for want of a set of Fleams, and a little Ingenuity. H.

Page 112.3. So about the House and Household Utensils, about the Barns, Stables, Pales, *etc.* twenty Things may be done by our Farmer and his Servants on rainy Days, and this (if it does not presently) will one Day turn to Account: however, at the present it turns to more Account than doing Nothing, or which is worse than Nothing, idling at the Ale-House. Yet this is not altogether our Author's Meaning, who would have your Barns against Harvest made tight, particularly from Drips (the most unknown of all Damages) all your Harvest-Tools ready and in good Order, and your Servants too: that when God sends you a good Crop, you may have nothing to do but to thank him, and rejoice like a Giant to run your Course. H.

Page 112.4. Woodsere is the Season of felling of Wood, as this Month is the properest Time to stub up what ye would destroy. The Heat of the Sun dries the Moisture of the Wounds very deep: and all Prunings at this Time dry further after the Knife, than at any other. So that with our Author, what you have a Mind to destroy, now cut it down, what you have not, let alone. H.

Page 112.5. Brambles, or common Bushes, may be now stub'd for Firing, where they annoy: but where they are wanted (as I take it at present in most Parts they are) namely, for fencing Wood, they are better let alone until fencing Time, both because then they are most wanted, and now they will be destroy'd, as in the foregoing Stanza. But this is the Time of the Year for Brakes, (if they are ready) which many Years they are not, until the next Month. Note, when you Mow these for Firing, the shortest and thickest are the best worth your while: for altho' a Man may mow two Load of long rank Blakes on his one that mows the short: yet after they are made and on the Cart, the Cart-load of small ones shall weigh one and a half of the great ones: and besides, shall lie in much less Compass, and rise in Flakes out of the Stack: As to the latter two Lines, every one knows when a Thing is full ripe, it improves no longer without altering its Condition. H.

Page 112.6. Of Head-Londs, or Hedge-Greens, has been spoken before, is the Time of cutting what is fit to cut. But why Grass upon Head-Londs of Barley, or Pease, should be let alone (until after Harvest) I cannot tell. It is true, they were sown much later than Winter Corn, but not so late that their Grass will not be fit to cut till after Harvest. However, since our Author concludes with, *Go mow if you please*, we may suppose every Man is left to his Liberty in this Case, and that the Reason why he put it beyond Harvest, was, because he thought it would not be fit before, and in Harvest the Mower might be better employ'd. H.

Page 112.7. Feying, is cleaning a Ditch or Pond, so as the Water may come clear. The Mus of these is excellent for mellowing stiff Ground, if mixt with Chalk: it is also excellent upon Pasture Ground, kindly refreshing the Root, especially for hot Gravely.

And altho' I find this was a Compost in our Author's Time, yet at present in Norfolk, I find nothing more disus'd: for as it mellows Clay, it would also stiff Sand. But Turnips I suppose supplies this, and many other Defects, which makes them less mindful of Composts, than their Neighbours of Cambridgeshire, Huntington and Bedford, who are most ingenious that way. H.

Page 113.1. Hops I take it were but newly come into Vogue in our Author's Time, for altho' they first began to be us'd in the Reign of King Henry the Eighth [about 1524], soon after his Expedition against Tournay; yet like other Improvements, they met with many ignorant Enemies [the physicians having represented that they were unwholesome, their use was prohibited in 1528, Haydn, *Dict. of Dates*]: however, the longer they were us'd, the better they were known: and at last many began to plant them, and amongst the rest our Author, and chuses his Ground as in the next. H.

Page 113.2. There is an Infancy due to all Inventions, which the Time our Author wrote in, I take to have been that of Hops, which are since much better known: however, this Rule holds still tolerably well: for altho' Grounds inclining to Sand, are found to produce good Hop-yards, yet too sandy is bad, and inclining to Clay, Stony or Rocky, wholly rejected at present. H.

Page 113.3 The Hop delights in the richest Land, a deep Mould and light, if mix'd with Sand it's the better. A black Garden-Mould is excellent for the Hop, says Mr. Worlidge, p. 145, *Systema Agricuturae*. The Hop delights most in rich black Garden-Mould that is deep and light, and that is mix'd rather with Sand than Clay, Mr. *Mortimer*, p. 132, *Art of Husbandry*.

If it, meaning the Hop Ground, lie near the Water, and may be laid dry, it is the better: M. *Worlidge*, p. 145.

So that modern Experience has not far out-gone our Author in the Judgment of what Ground is most fit, altho' Experience has taught us, that many Grounds that were formerly rejected, have since turn'd to very good account: for most Sort of Lands that are in good Heart, will do well enough, except as before excepted, the Stony, Rocky and stiff Clays. H.

Page 113.4. So that, as near as you can, your Ground must be open to the South, and fenced to the North and East. H.

Page 113.5. And therefore this Digression comes into this Month, for now is the scalding time to burn up the Roots of the Grass, and if it has been Meadow, now is its Crop of Hay off. H.

Page 113.6. There is, without doubt, a considerable Spirit in Hops, witness the Smell of Wort, when it first comes through (as the Brewers term beginning to boil) but this is for the most part lost in the Air, as being extremely volatile: However, there remains a Bitterness which is extremely grateful and digestive to the Stomach, and makes that keep and drink brisk, which otherwise would be both small and soure: keep, as our Author says, if it be drawn out its due length. H.

Page 114.11. "For wanting at will"—for fear of having none when you really want it. E.D.S.

Page 115.1. The Title of Captain is not at all here misapplied (altho' the Command be only over a Company of innocent Rustciks, whose Arms are Pitch-forks and Rakes, and their Ammunition Small Beer and Bread and Cheese) for here is required a due Prudence and Foresight, Celerity and Resolution, for it often happens one Hour well employed, may save the wasting of twenty: and if the Eye of the Master can make a Horse fat, it will make a Servant work. Mr. Trenchfield, in his *Cap of Gray Heirs*, etc., tells us a Story to this Purpose, of a certain Gentleman, who having wasted a great Part of his Estate by Mismanagement, sold the one half of it, and let the rest to a Farmer. The Farmer throve so well, that in a little time he offer'd to buy his Farm: This seem'd very strange to the Gentleman, who could not live upon twice as much of his own, as the other got an Estate out of, and paid Rent for. But the Farmer clear'd the Disproportion, by telling him, that the Difference lay in their frequent use of two Words only: You, said the Farmer, say *Go*, and I say *Come*: You bid your Servants go about this or that Work, and I say to my People, come Boys, let's go do this or that, *etc.*

Page 115.2. The *Norfolk* way of making Hay is, first to let it lie in the Swarth three Days, or more, then turn it: afterwards throw it into

Wind-rows, and thence cock it hot, and load it off as soon as they can. If it Mow-burn a little, they think of it ne'er the worse, for Neat Cattle will greedily eat, and it mellows the coarser Hay. But for Upland, or good Marsh-Ground either, this Way is not so good as that of Grass-Cocking, as it is used about *London*, and in these more Southern Parts: Here the Colour, Flavour, and true Sweetness is preserv'd: and tho' an Ox may be of another Mind, an Horse has Wit enough to thrive, work, or waste accordingly. *Note*, Mow-burnt Hay is very apt to breed the Bots in Horses. H.

Page 115.3. Tithes are of vast Antiquity, at least as old as *Abraham*, who paid Tithes to Melchizedec, *Heb.* 7. nay, it is not improbable, that the Offerings made by Cain and Abel, were first Fruits or Tenths: and it is naturally imprinted in the Mind of Man, that a Part of the Product of the Earth, ought to be dedicated to the Supreme Being, he who with his Rain and Sunshine produces it. As to the Abuses that have (by Man's deprav'd Nature) been made of such Dedications, they do not in the least countenance the Disuse of them, or any farther Abuses of them. H.

Page 115.4. *Avise else avous* is a Jargon for assure yourself, or be assur'd, Hay, if hous'd unmade, is of all Things the most apt to take Fire: what takes Wet by Rain, is not so apt to fire, but it turns to a filthy stinking Mouldiness, that nothing will touch. Coarse and long shady Hay is more coveted by a Cow or Ox, than the best hard Hay: for they having no upper Teeth, cannot chew it so well. Sheep are for the shortest Hay, and are somewhat more nice than Horses, and Horses, as before, love the best. H.

Page 115.5. The Hedlonds here meant, are the Hedge Greens formerly mention'd, which he advises to begin with: for here the Grass ought to be cut younger than in Meadows, because if it stand to Seed, it is apt to foul the arable Land. A Dallop is a Patch or Bit of Ground, lying here and there amongst Corn, which, either for its Moistness, Roots of Trees, or other Obstacle, has escaped the Plough: These our Author advises to seek out, and cut off their Grass, and bring it away Green, and make it elsewhere, to avoid its pestring the arable Land that surrounds it, with its Seeds, as it is very apt to do.

The Hedge Greens, about Barly and Pease, to be sure are thinnest, as having been fed down, and turned upon in the Spring, much later than those of the Wheat and Rye: yet if it is thin, it is better than nothing to carry off, and it is worse than nothing to stand, for the former Reason. H.

Page 115.6. Thry Fallowing is the third Plowing of a Summer Fallow, which he advises to be done betimes, that the Ground be a little hardned, before the Thistle and Dock seeds fly, that they may not take Root, but perish on the Ground.

He adds, indeed, that it may so happen, that you may be forc'd to plough it once more, before the Summer is ended: which if you do, you shall not lose your Trouble, but be paid for it in the next Crop: for the Pitch-fork in the Hay, the Shovel in the Malt, and the Plough in the Land, seldom go unrewarded. H.

Page 116.1. This is spoken of Garden Beans, which ought not be stript downwards, as some do: neither is it at present usual, or for the Gardner's Profit to cut them, but with a half Turn our Gardners at present twist them off: and this is perform'd much quicker and cleverer than cutting, and besides, fills the Bushel the sooner. H.

Page 116.2. See note to 106.2.

Fimble or Female Hemp, so called, I suppose, because it falls to the Females Share to *Tew Taw*, that is, dress it, and to spin it, *etc*. is the smaller, and when fit to gather, yellowish about the Stalk: It has a bended flower'd Head, not a knotted one, as the Carl Hemp (which is what bears the Seed) has: This, I suppose, is so called because it falls to the Carls or Churl's Share, our Author's *Michel*, and is very coarse, fit only for Cordage, *etc*. but its Seed makes amends, and bears near twice the Value of the other. H.

Page 116.3. Flax is often made a double Crop, namely Seed and Flax, but the Linnen is much better of such that is gather'd before it runs to Seed, being gather'd in the Bloom. It delights in a light rich Mould, and is a great Impairer of Land: therefore most proper to temper over-rank Grounds. Buck or Brank if now us'd to feed Cattle with upon the Ground, but no where to make a Sort of Hay of, as here our Author intimates. And it is very rare that it is ripe so soon as this Month, however, if it be, it lies abroad a good while after it is cut down, and altho' it suffers not much by Wet, yet it must be hous'd very dry, and if never so dry, there is no Fear of its shedding its Seed. H.

Page 116.4. Wormwood is certainly an Enemy to the Flea, but true hearty Cleanliness is a greater: for frequent washing a Room will prevent them, which is better than driving them out of one Room into another: howsoever, where a Room is infected with them, it must be rid of them, and this Way of our Author may do it well enough for ought I know. To get them out of a Bed, get good Store of Wormwood, lay it over your Mat or Ticking, over it lay a Blanket, and on it your Bed. After this Blanket smells well of the Wormwood, shift it from below, above you, and let all the rest go the same Circulation: be sure let your Bed be turn'd every Time it is made, and suffer no Dust about you, or as little as you can, for cleanly House-wives say, Dust breeds Fleas. H.

Page 116.5. Wormwood and Rue were in great Reputation in our Author's Time: and since him, we find Culpeper in his Herbal, has made a great Clutter about the Virtues of Wormwood: without Doubt they have their Virtues, but when too generally apply'd, as I am of Opinion here they are, they may sometimes do hurt as well as good: for Instance, Wormwood is found out to be an Enemy to the Nerves, and consequently to the Eyes. H.

Page 116.6. "As many doo more"—as many others do. E.D.S.
Against the approaching Harvest, Store of all Things should be laid in, especially Meal and Flower: that here be no running and fetching when the Work require all Hands, and if (as often it does about that Season) Water and Wind fail at the Mill, you will be sadly put to it indeed: Besides, your lying at the Miller's Mercy, who, in Harvest-time, for his fetching and carrying takes double Toll: and Millers are not bely'd when 'tis said of them, that they or their Servants have many crafty Tricks: one is this, on Pretence of haste of Work, they will set the Mill a going faster than ordinary, this shall heat the Meal so, that when it comes out, it shall suck in some much Moisture from the Air, as to be considerably heavier than it was before it was ground. H.

Page 121.1. Thry Fallowing is the third plowing of a Summer Fallow, which here he advises also to strike or harrow, to tear up the Weeds, especially the Couch Grass, by the Roots: and then to dung the same, for many Weeds, especially this of Couch Grass, will recover from a very small Root. H.

Page 121.2. Brakes (as I observ'd before) is their light Firing in Norfolk (that is that wherewith they Bake and Brew) these should be cut in dry Weather, or before the Rains come for two Reasons, first, that they may wither and be hous'd soon, and that the common Cattle may get at the Grass that grows under them, when the open Spaces are eaten bare. Why, June and August are the best Months to mow Brakes in: I take to be, because they are most usually mow'd in those Months, for they are extreamly tender in their Infancy, and a very small Frost when they first peep up will send them back again, so that when they are forward, and have receiv'd no such Check, they are fit for mowing in June, and when they are backward, namely after a frosty Spring, in August. H.

Page 121.3. The two St. *Mary's* here meant, I take to be the 22nd of *July*, being the Festival of St. Mary Magdelen, and the 15th of *August*, on which Day the *Roman* Church commemorate an Assumption of the Blessed Virgin. The Paring here spoken of, I take to be the taking up the Roots and transplanting them into fresh Ground, which our Author here advises to be at three Years End in the Field, and at four in the Garden: there is, however, Variety of Opinions in this Matter, some thinking March, some Midsummer the better Season. The Way of planting them is in Ranges made with a large Hoe, at four or five Inches Distance, and the Roots at two or three Inches Distance from each other. H.

Page 121.4. This agrees well enough with what may be done, for after the first Crop, Saffron makes a very good Sward, whereon Linnen may lye hollow and bleach well enough. H.

Page 121.5. Mustardseed is very apt to shed, and therefore must not stand until it is too ripe: it is best cut in a Morning when the Dew is yet on it, when dry house it with a Sheet carried between two, with a Pole on each Side: When strip'd (as our Author calls it) which is beating it upon a Hurdle or some other rough Thing, the Seed will come out: the light Seed will soon after appear white and thin, this must be well winnow'd off. H.

Page 121.6. This is meant of all Sorts of Garden Seeds, which our Author advises his Housewives to keep, and out of good Neighbourhood furnish one another with, for what greater comfort can there be than to be able to oblige with a little. Now if this is not practis'd so much as it ought to be at present, all that our Author did, or I can do, is to recommend it. H.

Page 122.1. Corn doth not only shed when it stands too long, but grows harsh, and loses much of its Beauty. If when God lays a Blessing before us, we neglect accepting it, we certainly are ungrateful: we should watch as well as pray. H.

Page 122.2. Our Author is justly against letting Harvest by the great, for whoever does, will certainly find himself cheated or slighted: he advises rather by the Day, but that is subject to great Inconveniences, if Men must be every Day look'd up. H.

[Though all the editions which I have seen read "By great will deceive thee," *etc.* I apprehend our author meant exactly the reverse. At least experience tells us that *day* labourers will deceive, while the *great* will despatch. In fact, in most operations of husbandry that can be reduced to a measure it is safest to let by the great, provided care be taken to see that the work is well done. In harvest, however, both sorts of labourers should be employed. The one for task work, the other to attend early or late, by agreement, to whatever business the season and circumstances may require. M.]

The best Way I take it, is what is now most in use, namely, to hire Men at Meat, Drink and Wages for the whole Harvest, then no Opportunity need be lost, and the Work will go roundly on. As to Provision (of which they will consume a great Quantity), by looking out in time it may be made easy enough, a Cow or two, some fatted Crones (old Ewes) may be timely provided, so as to go a good Way in your Family, and if you have but Plenty, and Fat, provided it be sweet, your Guests will ask no further Questions: for at this Time they do expect a full Diet, and he that keeps a plentiful

House, shall have more Servants at Command another Year, than he that gives a Crown more in Wages, and pinches, neither shall his Work be so well done. H.

Page 122.3. He that is the Lord of the Harvest, is generally some stay'd sober working Man, who understands all Sorts of Harvest-Work [*Cf.* Matt. ix, 38]. If he be of able Body, he commonly leads the Swarth in reaping and mowing. It is customary to give Gloves to Reapers, especially where the Wheat is thistly. As to crying a Largess, they need not be reminded of it in these our Days, whatever they were in our Author's Time. H.

Page 122.4. In this Stanza, in ten small Sentences our Author has describ'd all that is material in Harvest-Work, and of which (I think) there needs no Explanation, unless that a Gove is what in most Parts is call'd a Mow, which he advises to be kept true and upright, both for making the most of your Room, and keeping it from sliding. H.

Page 122.5. Of the Tithe somewhat has been spoke in former Months, therefore the less will serve here. It is certain the Tithe is not the Farmer's: and withholding it is Cheating, and Cheating never thrives. H.

Page 122.6. If the Parson is willing to have his Tithe justly paid, it is but Reason he should justly receive it, and not let it stand on the Ground to perplex the Farmer, who dare not bring in his Hogs or Cattel until it is taken away. H.

Page 122.7. This Guide is to take the fore Horse by the Head, and lead him straight in and may be done by the Boy or Girl who rake after the Cart. It is very proper to hinder overthrowing, and other Mischiefs. Hoying or hunting away the Hogs from under the Cart before it moves, is also very proper, lest the Wheel run over them. H. "Spight of thy hart"—to your regret.

Page 123.1. This is very often practis'd, for this Stubble if left long enough after the Sickle, is excellent good Thatch, very good light Firing for Brewing and Baking, and making of Malt: But the taking it thus away, impoverishes the Land, and where it is used, is a Sign of great Scarcity of Firing. H.

Page 123.2. Barley is at present most frequent mow'd altho' (in some of the Northern Parts) they continue to reap it, where Carts and Waggons are in use, it is set on Cocks, but where it must be carried on Horse Backs (as in Devonshire, or on Sledges, as in some Parts of Northumberland), it is bound up. H.

Page 123.3. Dallops are Tufts of Corn, such as are commonly seen where Dung Heaps have stood too long, or in shady Places: these he advises to let stand, and as occasion serves, cut them for Bands, where Bands are requir'd. Indeed these are commonly more empty ear'd, and if mix'd, apt to Mow-burn the rest, which they will not do when in Bands, and are besides most fit for that Use, by Reason of their Toughness and Length. H.

Page 123.5. This alludes to the Custom of Norfolk, where the Parson takes his Tithe in the Swarth, the Farmer also clears the Swarths, and afterwards with a Drag-Rake, rakes his Ground all over: it is true, the Tithe of this is as due as the other, but then the Parson ought to allow him for his Labour. H.

Page 123.6. Mow-burnt-Corn is easily known, for it is not only redder than ordinary in the Hand, but the very Flower or Inside is turn'd yellow, and is neither good for Bread-Corn, Seed nor Malt, as having spent its Fermentative Quality: neither is it good for Horses, because it breeds the Botts: and Poultry will scarce touch it, therefore ought to be avoided as much as possible. By well withering the Corn before it is hous'd: hous'd I say, because it is much more apt to Mow-burn in a House or Barn, than in a Stack: some prescribe leaving a Hole or Well in the Middle of the Mow, which may be done by keeping therein a Basket or Barrel, and raising it as the Mow increases, but no Remedy is so proper as the Prevention of the Disease. H.

Page 123.7. Pease ought to be turn'd a little before loaden, to dry that Side that hath lain next the Ground: and they of all Corn or Pulse contract most Moisture: But it does not follow they must not be turn'd until then, and indeed they require turning once if not more, or one half of them will go green into the Barn. H.

Page 124.1. It is best to keep many Goves or Mows going at the same time, that you may sort your Corn, and thresh that first that soonest needs threshing: your best Barley and best inned, being what you reserve for Seed, may lie farthest in, both because it comes last, and is out of the Malt Man's Reach, who, if he catches a Sample of it, will be apt to run down that which is worse. H.

Page 124.2. An old Sail is an excellent Thing for this Purpose: which may be laid over them all the Way they rise, and until you can thatch them. H.

Page 124.3. The Poor are the Sheep of God's Pasture, and therefore ought to be fed before the Farmer's: and this of gleaning, God was pleased to entitle them unto in the Levitical Law. But then these Poor must be the real poor, that is, such ancient People, Boys and Girls that cannot assist in Harvest Works, or at least that are not required, and I believe it is no Sin for a Farmer to turn that Gleaner out of his Ground, who is able and refuses other Work. After the Gleaner, come the Horses and Hogs, and after them our Author well advises, that it be kept up till after Michaelmas, that the Corn that is left on the Ground may sprout into Green. This is an excellent Food for Cows, and lengthens your Dairy: whereas if you let them in after the Gleaner, what Corn they lick up, serves but to dry them. H.

Page 124.4. In brewing for Harvest, and in Harvest, make three Sorts of Beer, the first Wort or Strongest, you may put by for your own Use, the second is what is called best Beer, whereof each Man ought to have a Pint in the Morning before he goes to Work, and as much at Night as soon as he comes in. If they work any thing extraordinary (as in Norfolk they often do, during the Moon-shine) their Share must be more: Small Beer they must also have Plenty in the Field. H.

Page 124.5. This the poor Labourer thinks crowns all, a good Supper must be provided, and every one who did any thing towards the Inning, must now have some Reward, as Ribbons, Laces, Rows of Pins to Boys and Girls, if never so small for their Encouragement, and to be sure Plumb-pudding. The Men must now have some better than best Drink, which with a little Tobacco, and their screaming for their Largesses, their Business will soon be done. H.

Page 124.7. For it is fitting they empty their Yards before they begin to thresh again their Cattle are in good Plight, and have little else to do, and they may go several Ways to their Lands, which they cannot at another time. H.

Page 125.1. Compass we know is Dung, now without doubt that which is rottenest is best and the sooner in the Field the better. H.

Page 125.2. There are a Sort of Wheels call'd Dredge Wheels, that in indifferent Weather will go over a Meadow without much hurting it: but they are heavy and low, and so load the Carriage, and therefore dry Weather is best both for your Ground and your Horses, especially if the Carriage be heavy, as Wood, Gravel, Timber, and Coal commonly is. H.

Page 125.3. This is to put Things, especially his Fire Wood, so about him, as to lie most convenient for his Use: what will pile, pile, what will not, lay it under the Wall of the House upright round some Tree or Pole, *etc*. H.

Page 125.4. This is a Winter Lodging for your Hog, who in the Summer requires cool and shade, but in the Winter time extreamly dreads the North and East Winds: from which this is not only a good Fence, but he has also the Warmth the Weather can afford. H.

Page 125.5. This Custom of picking out of the Sheaves all smutty Corn may be saved where the seed was well brin'd: for that takes off all the poor thin Corn which produces the smutty Ears: however, it may be worth the while to employ Children in picking it still, if it be but to take out the Cockle. H.

Page 125.6. Change of Seed is one of the best Pieces of Husbandry, and in divers Farms a Man may have Variety of Ground and good Change of his own. H.

Page 125.7. This Piece of Husbandry (except in some few Houses) is now out of Doors, the more is the Pity: but because I have spoke somewhat of it before, I shall only here add, that our Author was a sound Reformed, as may be seen by his Beliefe and other Works of his: yet neither did he nor the reform'd Church in his Days, reject the keeping of Lent, and Days of Abstinence, as Popish. There is a good Use as well as an Abuse to be made of them. H.

Page 125.8. By burnt to the Stone, I understand that such Fish as is dry'd on the Beach in too hot Weather, whereby it loses its Whiteness, and is apt to have a rank Smell, Garlickly some modestly call it, for Fish dries best in windy Weather. If packt in Pease Straw, it lies hollow from each other, and consequently keeps cool. H.

Page 126.1. I should think his Employment after his Dung is carried into the Field, should be to get his Winter Corn-Land ready: but if his Dung be upon his Land, it is best to spread it as soon as he can. H.

Page 126.2. This needs no Comment, and our Farmers now a Days know as well how to practise it as they did in our Author's Time: and who can blame them for endeavouring to make the best of what they have. H.

Page 126.3. Because it is the common practice of all Thieves: and two Horse-Stealers who live a hundred Miles from each other, shall chop and change their stolen Goods unpunish'd for a long Time. H.

Page 126.5. There are three Sorts of buying, in which there is a very great Difference: and indeed it is but reasonable there should be so, for besides the Interest, there is a very great Difference between running after a Debtor to get one's Money, and having it in one's Pocket and looking out for another Bargain. H.

Page 126.6. The Difference is, the Chapmen follow the ready Money Man, and they who go upon Trust, are fain to run after the Chapmen: and that makes good what our Author said in *Stanza* 41. That the rich Man is sought after both by Buyer and Seller. H.

Page 126.7. The Michaelmas Spring here meant, I take to be the freshning and managing your Pasture Ground so to your Advantage, that you may have wherewithal to keep your Catttle upon, as long as they will thrive upon it: of which there is a considerable Difference in Ground, particularly in low Grounds: some feeding much longer than others, he may also have regard in it to the sowing of Winter-Corn, for ought I know. H.

Page 126.8. The Market here spoken of, is in the Farmer's Travels mention'd before, which he advises not to be too long, and to drive home a Couple of Winter milch Cows, the one somewhat later than the other. These he may easily procure, for after Grass is gone, a Winter Milch Cow is enough to ruin a poor Man. H.

Page 127.1. Crones, I have said before are such Ewes whose Teeth are worn down, so that they can no longer live in their Sheep Walk, these are sometimes not very old, and when put into good Pasture will thrive exceedingly, and are at this Time often sold very cheap. I have known good ones at 1s. 10d. a Piece, with each a Lamb in her Belly: and these pay their Lamb, their Fleece and their Flesh, for their Food before Harvest next. It is now a good Time to dry up your old Cattle, and with Care they will be tolerable good Christmas Beef. H.

Page 127.2. This alludes to Norfolk, Suffolk and Essex, where this Fair and some others stock the Country with Clothes, and all other Household Necessaries: and they again, sell their Butter and Cheese, and whatever else remains on their Hands: nay, there the Shopkeepers supply themselves with divers Sorts of Commodities. H.

Page 127.3. The Colour of the Hop, is that which makes it valuable in the Eyes of a great many People, and indeed a glorious Colour is a Beauty: however, a little of it may be abated, provided it be made up with innate Goodness. Now, a Hop a little brownish, has not lost much, nay, is often better than the over-bright: however, there is a mean, which our Brewers know very well how to chuse. H.

Page 127.4. I take this Caution to be of no great Value: for Hops are more easily cut, than broken off, especially when on the Pole. The paring the Hill about, and turning the Grass inwards, cherishes and arms the Root against the ensuing Cold, and is of very good Use. H.

Page 127.6. This is for gathering them, which he advises to be without breaking the Poles and then directs them to be picked, either upon Soutage, which is the Cloth, they are generally packt in, or the Hair Cloth that covers the Kiln. There are a Sort of Troughs now much in Use, better than either. H.

Page 127.7. Ash, Beech and Birch, and some Oak too, are now frequently used for Poles. H.

Page 127.8. Kell-drying is without Doubt the most practicable Way because done at a Certainty, and may be made ready to any Market in View. But for small Quantities, Soller or Garret drying may do very well. H.

Page 128.1. And I have only thus much to shew, namely, that if Hops are at Seven Pounds the Hundred, your Soutage stands you in one Shilling and three Pence the Pound, and if they us'd to pack in Canvas in our Author's Days, methinks they might also now, when a Price will afford it: for the closer they are packt, the longer they keep their Strength: and therefore in some Cases they may be put up in Cask, especially for private use. H.

Page 132.6. "To him and to hur"—to everyone, or to anyone. E.D.S.

Page 133.7. "Tode with a R"—an elegant euphuism for *torde;* the meaning being that a bad husbandman is more likely to receive insults and refusals than compliance with his requests. E.D.S.

Page 134.2. This is spoken of Champion, or open Field Land, Ironically calling these the Fences to the Meadow and Corn, which are the greatest Nusances. A Balk, is what in some Places is call'd a mier Bank, being narrow Slips of Land between Ground and Ground. H.

Page 134.3. The Feed is commonly swept all at once, and the Sheep will be down before their Time. H.

Page 135.3. There is very good Cheese as well as bad made in Suffolk, but the great Dairies starve their Cheese for their Butter. Prest is an Old Word, for Neat or Tight, I suppose comes from Women being strait-lac'd. H.

Page 135.5. In Norfolk (in our Author's Time) there was a considerable Rebellion, call'd Ket's Rebellion [1549] against Inclosures, and to this Day they take the Liberty of throwing open all Inclosures out of the common Field, these are commonly call'd Lammas Lands, and half Year Lands. H.

Page 136.1. Field Gates cannot always be kept shut, great Roads frequently lying through them, and then especially when the Commons are bare, common Cattle are apt to throng in. Where they border upon Warrens, Conies will run a great Way into them. Conies are best fenc'd out by observing their Haunts, and thrusting Bushes, Brambles, or Furzes into them, also topping the Hedge with Furzes, so as that they may hang over, is a good way, but a wet Ditch if possible to be had, is the best Fence: Against Swine there is scarce any Fence, except a Wall or Pale: a Dog to follow or shake him by the Ear is somewhat, but there is much Corn broke down by their running. The best way is for every one to agree to keep them up, when there is nothing to be got by them abroad but what they steal. H.

Page 136.2. In ancient Times their Winter Corn was not so soon in the Ground as in nearer: and in many Courts the Limitation of the Flocks feeding is much longer, than not only our present Improvement of Husbandry, but that of our Author's Time would allow. H.

Page 136.3. There are a great many such Towns at present, but the more is the Pity: for indeed here lies the whole Grievance, and because of Perjury the Nation justly mourns. H.

Courts for presenting nuisances are generally the greatest nuisances themselves. Under the semblance of justice they often retard its execution. The members or jury who compose them do not want the power, but they want the independance to act right. M.

Page 137.2. "Take them"—arrest them. E.D.S.

Page 138.2. Here are enumerated Abundance of Inconveniences that Champion Land undergoes, in comparison to Enclosed, and all very true: for where there is a great deal, what is every Body's Care, is no Body's Care: for it is not only the Shepherd, the Ox-boy and the Poor, but Farmers and Gentlemen will filch from one another, form pretended Privileges out of bad Customs, such as Foot-paths, Sheep-drifts, Privilege of Hunting and Hawking: in all which, they shall frequently do Mischief out of Malice, as well as Covetousness. The Foot-path was at first conniv'd for the Conveniency of some new built House, or the like: this soon becomes a Horse-way, and in a little time a Road. The Sheep-way perhaps at first, went all thro' the Sheep-Owner's Land, or some untill'd Space. In Process of Time the Farms are otherwise divided, and this Ground becomes good

Arable, and is in Tilth: upon any Spite the Sheep shall go through it still, and the Crop shall be eaten to the Ground, and the best Remedy for the Injur'd, namely, A Suit in Law turns to the worst Account. The Lord preserves a Privilege of Hunting and Hawking, and with this Privilege he shall vaunt and insult his richer and more careful Neighbour, nay, and endamage him too at his Pleasure: and if he sues for Remedy, our Law allows him no more Cost than Damage: These are in a great measure remedied by Inclosures, the Stile hinders the Path from becoming a Horse-way, and the Hedges on both Sides keep the Sheep within their Bounds, and the Gallant is probably now more afraid of his Neck, than before he was of his Neighbour's Livelihood. H.

Page 138.3. I remember, I saw a Man once throwing in some Pease pretty late in the Evening, How now Neighbour, said I, you are late at work, Ay, ay, replied he, Field-land, Field-land, one can call noting one's own, until it is in the Barn. And he said true: for next Morning I saw he had thrown a Land's Breadth of mine into his: Now, whether he did it out of Knavery or Ignorance, matters not, it could not have been done in an Enclosure, and those who have experienced it, know what mad Work a high Wind will make amongst Pease and Barley Cocks in a common Field, when in an Enclosure the Hedge stops all. H.

Page 139.2. It is likely this was wrote soon after Ket's Rebellion, as a Dissuasive from the like, and to persuade the poorer Sort quietly to endure the Enclosures, which certainly are more beneficial in the main to the Poor, than all their pretended Privileges: for where there are Enclosures, there is a constant Succession of Work: whereas in Champion, Harvest and Threshing is almost all they have. H.

Page 139.3. The Wood Lands is that Part of Norfolk, which lies about Watton, Hingham and East Dereham, where indeed are very pretty Habitations: and where I think every thing looks much more chearful than any other Part of that Country: But here may be taken in general. It is true, two Acres of Enclosure is but a very poor Man's Farm, no more is twenty of Arable, especially if a poor Team must be kept to plough it: however, that this two Acres of Meadow or

Pasture enclos'd, and near a good Common, shall clear more at the Year's End, than the twenty of Champion: is plain to whosoever will consider. The two Acres is only for Hay to winter, and after Grass to succour a Cow or two, or perhaps a few Ewes and Lambs, and all the poor man's Time is sav'd for Day Labour, whereas, the others is most, if not all laid out upon his Team and his Land. H.

Page 139.5. Our Author closes with a Truth which we see daily practis'd, and which I believe was in Use in his Days, as well as ours: that is, that the Rich shall share the Common amongst themselves, and let the Poor have no Proportional with them: nay, what remains after Encroachment, shall be the more swept with the rich Men's Stock, who now lies more convenient for it than before: This is enough to make a poor Man grutch, because he has but a little, he shall have less: and (as in all the Insurrections and Rebellions we read of) we find none to consist of so mean People, and none so stout and obstinate as Ket's Rebellion: I am apt to believe they had some Provocations from the Gentry, against whom their particular Bent was. In short, as the Common is not the Poor's, as Poor, yet according to the Freehold they rent or enjoy, they have a Share in every Division or Encroachment, and altho' no Encroachment will justify the flying into a Rebellion, it will justify a Complaint, and Desire to be reliev'd, and the taking all lawful Opportunities to be so. H.

Page 140.20. Davus is the common name in Terence for the cunning plotting servant. E.D.S.

Page 140.21. Thersites, the ugliest and most scurrilous of the Greeks before Troy, is used to denote a calumniator. E.D.S.

Page 143.1. At Canterbury is a representation of Master Shorne holding up his hand in a threatening attitude at the Devil, who is in a boot. E.D.S.

Page 143.2. "False birds can fetch the wind"—An expression taken from hawking, meaning to gain an advantage. E.D.S.

Page 146.4. "Could the way to thrive." *Could* is used in its old sense of *knew* or *understood*. E.D.S.

Page 147.3. "The blacke oxe neare trod on thy fut" A proverbial expression, meaning you have experienced good fortune close at home. E.D.S.

Page 154. William, the first Lord Paget, and the patron of Tusser, married Anne, daughter of Mr. Prestin, of the county of Lancaster, and to her, it is most probable, the Book of Huswifery was dedicated, and not to Margaret, the daughter of Sir H. Newton and lady of Thomas, Lord Paget. M.

Page 156.6. From the last stanza it would appear that the author was a widower when he wrote it. M.

Page 158. This antithetical description seems to have been introduced in order that it might correspond with the Description of Husbandry, Chap. 8. M.

Page 171.2. Though all the standard editions read, "chaps walking", may it not be a misprint for *chaps wagging*, *i.e.* mouths craving? M.

Page 177.4. The Skreene was a wooden settee or settle, with a high back sufficient to screen the sitters from the outward air, and was in the time of our ancestors an invariable article of furniture near all kitchen fires, and is still seen in the kitchens of many of our old farm houses in Cheshire. E.D.S

Page 179.1. "*Aqua Composita.*" A recipe is given in Cogan's Haven of Health (ed. 1612): "Take of Sage, Hysope, Rosemarie, Mynt, Spike or Lavender leaves, Marjoram. Bay leaves, of each like much, of all foure good handfulles to one galon of liquour. Take also of Cloves, Mace, Nitmegs, Ginger, Cinnamon, Pepper, Graines of each a quarter of an ounce, Liquorice and Annise, of each halfe a pound: beat the spices grosse [coarse] and first wash the herbes, then breake them gently betweene your hands. Scrape off the barke from the Liquorice, and cut it into thin slices, and punne [beat, pound] the Annise grosse, then put altogether into a gallon or more of good ale or Wine, and let them steep all night close covered in some vessell of earth or wood, and the next morning after distill them with a Limbecke or Serpentine. But see that your fire be temperate, and that the head of your Limbecke be kept colde continually with fresh water, and that the bottom of your Limbecke be fast luted with Rye dough, that so Ayre issue out. The best Ale to make Aqua Composita of is to be made of Wheat malte, and the next of cleane Barley malte: and the best Wine for that purpose is Sacke."

Page 182.6. As a practical musician himself, Tusser seems to consider music, for such as have an ear, as a proper branch of education. For boys it would be degrading, except as a profession: and even for girls it is generally worse than useless as it occupies that time which ought to be devoted to more important purposes. M.

Page 182.7. "Least a homelie breaker"—lest in inexperienced teacher ruin the mind of the pupil. E.D.S.

Page 183.1. "Well a fine"—to a good purpose, to a good result. E.D.S.

Page 184.5. "To purchase Lynn" by petty savings seems to have been a proverbial mode of expression used in ridicule of stinginesss. M.

Page 184.7. "Needham's shore" is a proverb quoted by Ray, signifying that waste and extravagance bring a man to want. M.

Page 190. "Still presently"—always as close at hand. E.D.S.

Page 202.2. "The author means London; but though it is believed he died there, it is evident from the sequel that he left it on account of the plague." M.

Page 206.2. "Norfolk wiles"—"Essex miles, Suffolk stiles, Norfolk wiles, many men beguiles"—*old East Anglian saw*. The "wiles", that is, the litigious propensities of the Norfolk people, are referred to by Fuller in his *Worthies*.

Page 206.3. "A Moone of cheerfull Hew" is a reference to his second wife. E.D.S.

APPENDIX

INTRODUCTION BY SIR WALTER SCOTT
AS REPRINTED IN THE 1931 TREGASKIS EDITION

THOMAS TUSSER, the author of this excellent and curious work, has given us a short abridgment of the principal events of his life. He was born about 1523, according to Ellis, "of lineage good, of gentle, blood," at a village called Rivenhall, in Essex, and having apparently a fine voice, was sent to study music at the collegiate chapel of Wallingford. The superintendants of choirs had at that time a sort of impress warrants, by which the boys educated to music were transferred to their different establishments. Our author had the good fortune to be sent to St. Paul's, where he attained considerable proficiency in music, under the tuition of John Redford. The next stage of his education was Eton school, where the discipline of Nicholas Udal seems to have made a more lasting impression on his mind than the classical lessons which accompanied it. A few scraps of Latin, and his version of St. Bernard's verses, with the classical allusions scattered through his poems, do however shew, that the fifty-three stripes, all bestowed upon him at one time, were not entirely thrown away.

Tusser then entered Trinity Hall at Cambridge, where his studies being interrupted by sickness, he became induced to engage in the service of Lord Paget, to which he was probably preferred by his knowledge of music. With this generous patron he remained in early servitude for ten years. The discord among the nobility, which began to arise in the reign of Edward VI drove Tusser, a man of quiet and pacific habits, from attendance upon Lord Paget, afraid peradventure his services might be required in a military capacity. He then addicted himself to husbandry, married and settled in Suffolk, upon the sea coast, but the air disagreeing with his wife's infirm constitution, he removed to Ipswich, where she died. We next find Tusser at Ratwood, in Suffolk, where he devised the plan of the following work. About the same time he was again married to Mistris Amy Moone, and soon afterwards removed his residence to Dixam, where he held a farm under Sir Richard Southwell.

The death of this gentleman, whose estate fell to seven executors, exposed Tusser to new difficulties. The quarrels betwixt these joint proprietors, whom he likens, not unaptly, to ravens, drove him from his farm to reside in the town of Norwich, where he became known to Salisburie, Dean of Norwich, was a singing man in the cathedral there, and, probably through the dean's influence, obtained a lease of tithes near Fairsted, in Essex. The lease however being only for the parson's life, and the occupation subject to much loss and vexation, our author again renounced his situation and went to London. From London he was driven by the plague to Cambridge, where he obtained some situation in his college, and moved, doubtless by the classic air which he breathed, revised and published his Georgics, or, as he unostentatiously calls them, "Five Hundred Points of Good Husbandry."

Apparently, however, he afterwards quitted his haven at Cambridge, to adventure forth again into a world which he had found so unstable, for he died very aged, in London 1580, and was buried in St. Mildred's church, in the Poultry, where his grave bears the following inscription, composed by himself, or an equally homely Muse

> *Here Thomas Tusser,*
> *Clad in earth, doth lie,*
> *That sometime made*
> *The Points of good Husbandrie.*
> *By him then learn thou mayest,*
> *Here learn thou must,*
> *When all is done we sleep,*
> *And turn to dust.*
> *And yet, through Christ,*
> *To heaven we hope to go:*
> *Who reades his bookes,*
> *Shall find his faith was so.*

It is obvious from the incidents of his life, that, notwithstanding his excellent maxims of frugality, he did not himself profit by his agriculture or economy. His ill success occasioned the following epigram:—

> *Tusser, they tell me, when thou wert alive,*
> *Thou teaching thrift, thyself could never thrive,*
> *So like the whetstone, many men are wont,*
> *To sharpen others, when themselves are blunt.*

Wit's Recreations 1641

"He was successively," says Fuller, "a musician, schoolmaster, serving-man, husbandman, grazier, poet, more skilful in all than thriving in any profession. He traded at large in oxen, sheep, dairies, grain of all kinds, to no profit. Whether he bought, or sold, he lost, and, when a renter, impoverished himself, and never enriched his landlord. Yet hath he laid down excellent rules in his book of husbandry and houswifry, (so that the observer thereof must be rich) in his own defence. He spread his bread with all sorts of butter, yet none would stick thereon. Yet I hear no man to charge him with any vicious extravagancy, or visible carelessness, imputing his ill success to some occult cause in God's counsel. Thus our English Columella might say with the poet,

Monitis sum minor ipse meis,

none being better at the theory, or worse at the practice, of husbandry. I match him with Thomas Churchyard, they being marked alike in their poetical parts, living at the same time, and statured alike in their estates, being low enough, I assure you."

But however Tusser may have failed in setting an example, the excellence of his precepts has never been disputed. There is no where to be found, excepting, perhaps, in Swift's Directions to Servants, evidence of such rigid and minute attention to every department of domestic economy: and if Tusser's observations are less entertaining, than those of the Dean of St. Patrick's, it must be remembered they are compiled for use, not for satire. If indeed his genius had been that way directed, his said remembrance could not but have stocked him with numerous examples of the sloth, fraud, and waste of domestics. For although he was able to lay down the strictest rules of economy, for feeding every living thing within his gates, from the master's hall to the ban-dog's kennel, the extravagance of his servants was the principal cause which disgusted him with the life of a husbandman.

Loiterers I kept so many,
Both Philip, Hob, and Cheanie,
That, that way nothing geanie,
    Was thought to make me thrive.

Tusser's practical rules of husbandry have been thought excellent by unquestionable judges. Lord Molesworth proposed, that, to increase the number of husbandmen, and prevent the growth of the idle poor, the book of husbandry should be taught at parish schools, as a sort of manual both of knowledge and moral precept. In the former point of view, many of the lessons may have become obsolete. Yet these remain interesting to the agricultural antiquary, and not to him alone, but to all who are curious to know the simple, orderly, and strictly economical mode of life of the English farmers in the 16th century.

Many old customs may be traced in Tusser's rude poetry, and some curious inferences drawn respecting the state of the peasantry. The English farm-servants lived better even in 1557, than the farmers themselves in Scotland, or on the continent do at this day. They looked, "of custom and right", to have roast meat on Sundays and Thursdays, and had besides various days of festival, to be regularly kept at the farmer's expence. Yet the patriarchal government under which they lived authorised occasionally a good drubbing. The maids in particular were subjected to this domestic discipline, as we learn from numerous hints throughout the work. As, for example,

> *Let Holly-wand threat,*
> *Let Fizgigg' be beat.*

Warton enumerates the following particulars of information concerning the farmer's domestic life, extracted from Tusser, to which many more might be added: "For the farmer's general diet he assigns, in Lent, red herrings and salt fish, which may remain in store when Lent is past: at Easter, veal and bacon: at Martinmes, salted beef, when dainties are not to be had in the country: at Middsummer, when mackrel are no longer in season, grasse, or sallads, fresh beef and pease: at the Michaelmas, fresh herrings, with fatted crones, or sheep: at All-Saints, pork and pease, sprats and spurlings: at Christmas, good cheere and plaie. The farmer's weekly fish-days, are Wednesday, Friday, and Saturday: and he is charged to be careful in keeping embrings and fast days. Among the husbandlie furniture, are recited most of the instruments now in use, with several obsolete and unintelligible names of farming utensils. Horses, I know not from what superstition, are to be annually blooded on Saint Stephen's day. Among the Christmas husbandlie fare, our author recommends good drinke, a good fire in the hall, brawne, pudding and souce, and mustard withall, beef, mutton, and pork, shred, or minced pies of the best, pig, veal, goose, capon, and turkey, cheese, apples, and nuts, with jolie carols. A Christmas carol is then introduced, to the tune of King Solomon.

"In a comparison between champion and several, that is, open and inclosed land, the disputes about inclosures appear to have been as violent as at present. Among huswifelie admonitions, which are not particularly addressed to the farmer, he advises three dishes at dinner, which being well dressed, will be sufficient to please your friend, and will become your hall. The prudent housewife is directed to make her own tallow candles. Servants of both sexes are ordered to go to bed at ten in the summer, and nine in the winter: to rise at five in the winter and four in the summer. The ploughmen's feasting days, or holydays, are Plough-Monday, or the first Monday after twelfth-day, when ploughing begins in Leicestershire. Shrof-tide, or Shrove-Tuesday, in Essex and Suffolk, when, after shroving or confession, he is permitted to go thresh the fat hen, and if blindfold (you) can kill her, then give it thy men, and to dine on fritters and pancakes. Sheepshearing, which is celebrated in Northamptonshire with fritters and cakes. The Wake day, or the vigil of the church saint, when everie wanton maie danse at her will, as in Leicestershire, and the oven is to be filled with fawnes. Harvest-home, when the harvest-home goose is to be killed. Seed-cake, or a festival so called at the end of wheat sowing in Essex and Suffolk, when the village is to be treated with seed-cakes, pasties, and the frumentie-pot. But twice a week, according to ancient right and custom, the farmer is to give roast-meat, that is, on Sundays and on Thursday-Nights." The few notes scattered through the tract will point out other circumstaces worthy of the antiquary's notice, although the present editor's experience does not enable him to throw any light upon the agricultural precepts or practice.

To inlist as many favourers of Tusser as possible, the merry hunter may be pleased with the moderation of the following precept:

To hunters and hawkers, take heed what you say,
Mild answer with courtesy drives them away:

> *So where a man's better will open a gap,*
> *Resist not with rudeness for fear of mishap.*

The poetry of Tusser is obviously the least recommendation of his work. Yet, even here, there is something worthy of observation: for the stanza of the following apology forms the first example of that employed by Shenstone in his pastoral ballad:

> *What lookest thou herein to have,*
> *Fine verses thy fancy to please,*
> *Of many my betters that crave,*
> *Look nothing but rudeness in these.*
>
> *What look ye, I pray ye shew what,*
> *Termes painted with rhetoric fine,*
> *Good husbandry seeketh not that,*
> *Nor is't any meaning of mine.*

Although neither beauty of description, or elegance of diction, was Tusser's object, he has frequently attained what better indeed suited to his purpose, a sort of homely, pointed, and quaint expression, like that of the old English proverb, which the rhyme and the alliteration tended to fix in the memory of the reader. To attain this concise and magisterial brevity of expression, he almost always discards articles, conjunctions, and even auxiliary verbs, where the sense can be attained without their assistance, and these frequent elisions constitute the peculiarity of his versification. The moral reflections of Tusser assume the quaint old-fashioned appearance of his agricultural instructions: and though just and lively, have no claim to elegance or sublimity.

In transferring this first specimen of English Didactic poetry to this collection, the editor has followed the edition of Short in 1599. The following apt motto may be adopted on Warton's suggestion:

> *Possum multu tibi veterum praecepta referre*
> *Ni refugis tenuesque piget cognoscere curas*
>
> Georgic, 1. 176.

GLOSSARY

ADDLE, *v.* yield, ripen fruit.
ADVISE, *n.* care, notice. "Take advice of thy rent"—make preparations for paying your rent.
AFTER CLAPS, *n.pl.* disagreeable and unexpected consequences.
AFTER CROP, *v.* extract a second crop from the land.
ALEXANDERS, *n.* (*Smyrnium olusatrum*).
ALL IN ALL, the principal point.
ALLEY, *n.* path, walk.
ALLOW, ALOW, *v.* recommend, approve of.
ALOWE, *adv.* low down, deep.
AMONG, *adv.* at times.
ANDREW, St. Andrew's Day, 30th November.
ANOIENG, *v.* injuring, damaging.
APERNE, *n.* apron.
ARAID, *pp.* kept in order, regulated.
ARMER, *n.* help, assistance.
ASSAI, *n.* trial.
ATCHIVE, *v.* finish, complete.
ATHIT, EVIL ATHIT, ill-conditioned, inferior, in the extreme, (? at height).
A TOO, *adv.* in two, asunder.
AUKE, *adj.* unlucky, untoward.
AUMBRIE, *n.* cupboard, pantry.
AVAILES, *v.* is useful or profitable.
AVENS, *n.* herb bennet (*Geum urbanum*).
AVISE ELSE AVOUSE, look out for, expect.
AVOUCH, *v.* own, acknowledge.
AWE, *n.* August.
AYER, *n.* air.

BAGGAGE, *n.* foul stuff.
BAGGEDGLIE, *adj.* worthless, rubbishy.
BAIES, *n.* chidings, reproof.
BAILIE, *n.* bailiff, steward
BAITING, *n.* feeding, eating.
BALKE, *n.* narrow slips of land between ground and ground.

BALL, *n.* horse.
BAND, *n.* band or rope of straw.
BANDES, *n.* bonds, engagements.
BANDOG, *n.* dog always tied up on account of fierceness.
BANE, *n.* ruin.
BANISH, *v.* free, clear.
BANKET, *v.* feast, banquet.
BARBERIES, *n.* fruit of the barberry (*Berberis vulgaris*).
BARBERLIE, *adv.* like a barber.
BARELIE, *n.* barley.
BARGAINE, *n.* contract, agreement.
BARTH, *n.* warm sheltered place for livestock.
BARTILMEWTIDE, St. Bartholomew's day, 24th August.
BASSEL, BAZELL, *n.* the kitchen herb basil (*Ocymum basilicum*).
BAULME, *n.* Balm (*Melissa officinalis*).
BAVEN, *n.* bundle of brushwood.
BEARE OFF, *v.* ward off, keep off.
BEARE OUT, *v.* keep off, protect from.
BEARES, *v.* provides, furnishes.
BEASE, *n.* beast, livestock.
BEASTLIE, *adj.* stupid, careless.
BEATH, *v.* warm unseasoned branch wood at the hearth to straighten it.
BECK, *n.* beak.
BECLIP, *v.* anticipate, surprise.
BEDSTRAW, *n.* clean straw.
BEENE, *n.* property, wealth.
BEERE, *n.* bier.
BEETLE, *n.* a wooden mallet.
BEGON, *pp.* begun.
BEHOOVING, *adj.* belonging, proper to.
BESHREAWD, *pp.* ruined, cursed.
BESTAD, *pp.* situated.
BESTOWE, *v.* place, arrange.
BETANIE, *n.* betony (*Stachys officinalis*), common wild plant once much valued in medicine.
BEWRAIES, *v.* betrays.

321

BEX, *n.* beaks.
BIG, *n.* teat, pap.
BIL, BILL, *n.* billhook.
BILLET, *n.* chopped-up wood.
BIN, *pp.* been.
BLABS, *n.* talkative persons.
BLADE, *n.* blade of grass.
BLAZE, *v.* spread abroad the report of.
BLEETS, *n.* pot-herb, (? Good King Henry, *Chenopodium bonus-henricus*).
BLENGE, *v.* blend, mix.
BLESSED THISTLE, *n.* milk thistle (*Silybum marianum*) veined and spotted with white, traditionally from the Virgin Mary's milk.
BLINDFILD, *adj.* blindfold.
BLISSE, *v.* bless, praise.
BLOODWORT, *n.* red-veined dock (*Rumex sanguineus*).
BLOUSE, *n.* red-faced wife or girl.
BLOWNE, *pp.* reported.
BOBBED, *pp.* pouting.
BODDLE, *n.* corn marigold (*Chrysanthemum segetum*).
BOLD, *v.* embolden, encourage.
BOLD, *adj.* proud.
BOLL, *n.* washing-bowl, tub.
BOLTED, *pp.* sifted, examined.
  Bolted-bread—a loaf of sifted wheat meal mixed with rye.
BOOLLESSE, *n.* bullace, fruit of *Prunus spinosa institia*).
BOORD, *n.* boards, planks.
BOORDE, *n.* table, meals.
BOROUGH, *n.* burrow, warren.
BOTCH, *v.* patch.
BOTLES, *n.* corn marigold (*Chrysanthemum segetum*).
BOTS, *n.* disease (worms) troublesome to horses.
BOTTLE, *n.* leather bottle.
BOWD, *n.* weevil.
BOWS, *n.* boughs, sticks.
BRAG,, *n.* boast, sham, pretence, value, estimation.
BRANK, *n.* buckwheat.
BRAVE, *adj.* fine, grand.

BRAVERIE, *n.* show, boast.
BRAWNE, *n.* flesh, meat.
BRAWNETH, *v.* fatteneth.
BREACHING, *n.* breaking, breach.
BREADCORNE, *n.* corn for making bread.
BREAKER, *n.* horse-breaker.
BREATHELY, *adj.* worthless.
BRECKE, *n.* breach, gap.
BREERS, *n.* briars, thorns, hence troubles and difficulties.
BREMBLE, *n.* bramble, briar.
BREST, *v.* nourish.
BREST, *n.* voice.
BRIBING, *v.* thieving, stealing.
BROTHELL, *v.* dissipate.
BRUSH, *n.* underwood, brushwood.
BRUSHED COTE, a beating.
BUCK, *n.* buck-wheat.
BUCKLE, *v.* prepare, get ready.
BUCKS, *n.*—washes, collections of dirty linen for washing.
BUGLAS, *n.* bugloss (*Lycopsis arvensis*).
BUIE, BUIENG, *v.* buy.
BULCHIN, *n.* bull-calf.
BULLIMONG, *n.* dredge, mixture of grains and seeds grown for cattle food.
BURRAGE, *n.* borage.
BUSHETS, *n.* small shoots from bushes.
BUSHT, *adj.* thick, spreading.
BUTTRICE, *n.* farrier's knife for paring horse hooves.
BUTTRIE, *n.* pantry, cupboard.

CABBEN, *n.* house, sty.
CADOW, *n.* jackdaw.
CAMPE, *v.* play football.
CAMPERS, *n.* football players.
CARELES, *adj.* unwilling, not anxious.
CARKE, *v.* be anxious.
CARNELS, *n.* hips and haws.
CARREGE, *n.* carrying home.
CARREN, *n.* carrion, carcases.
CARRENLY, *adj.* rotting, putrifying.
CAST, *v.* clean the threshed corn by casting it from one side of the barn to

the other, that the light grains and dust may fall out.

CAST, *v*. give over, throw up.

CATER, *n*. caterer, provider.

CAWME, *adj*. calm, settled.

CHALLENGE, *v*. claim.

CHAMPION, *n*. plain open country.

CHAMPIONS, *n*. inhabitants of countries where lands are open and unenclosed.

CHANCING, *v*. happening, falling out.

CHARGE, *n*. trouble, expense.

CHARGED, *pp*. burdened, busy, anxious.

CHARGES, *n*. works, troubles.

CHARVIEL, *n*. chervil (*Anthriscus cerefolium*).

CHAUNTING, *v*. crying, yelling.

CHEANIE, Jeanie, Jennie.

CHEERE, *v*. enjoy oneself.

CHEIN, *n*. chain.

CHIPPINGS, *n*. fragments of bread.

CHOPPING, *v*. exchange, barter.

CLAP, *n*. blow, stroke: "at a clap"—at once.

CLAPPER, *n*. rabbit burrow or warren.

CLARIE, *n*. clary (*Salvia sclarea*).

CLAVESTOCK, *n*. chopper for splitting wood.

CLICKET, *v*. chatter.

CLIM, ? short for Clement, as if addressing a worker so named.

CLOD, *n*. earth, hence—landed property.

CLOG, *n*. charge, duty.

CLOSET, *n*. retirement, seclusion.

CLOSETH, *v*. incloses, fences in.

CLOT, *n*. clods.

CLOUGHTED, *pp*. See CLOUTED.

CLOUT, *n*. piece of cloth.

CLOUTED, *pp*. fixed with iron plates.

CLOUTS; CLOUTES, *n*. iron plates fixed to a plough.

COAST, *n*. country, district.

COAST MAN, *n*. masters of coasting vessels.

COCK, *v*. put into cocks, or small stacks.

COCKING, *adj*. over-indulgent.

COCKLE, *n*. corn cockle (*Agrostemma githago*).

COCKNEIES, *n*. spoilt or effeminate boys.

CODWARE, *n*. all plants that bear pods (or cods): peas, beans, etc., pulse.

COEME; COOME, *n*. measure of half a quarter, i.e. four bushels.

COG, *v*. cheat, defraud.

COILE, *n*. bustle, hard work.

COLD, *adj*. cooling.

COLE, *n*. turf, peat.

COLEWORT, *n*. or collet, cabbage.

COLLECTS, *n*. prayers.

COLLEMBINES, *n*. columbine.

COMFORT, *n*. strength, fertility.

COMMODITIES, *n*. advantages.

COMPACT, *pp*. composed.

COMPAS, *n*. manure, compost.

COMPOUND, *v*. agree, arrange.

CONFER, *v*. compare.

CONFOUND, *v*. destroy, spoil.

CONIE, *n*. rabbit, "bunny", as a term of endearment for girls.

CONIES, *n*. rabbits.

CONSTANCIE, *n*. consistency, firmness.

CONSTER, *v*. understand.

CONTEEMNE, *v*. despise.

CONTINUE, *v*. to breed from, to keep up stock from.

COOSEN, *v*. cheat, swindle.

COPIE, *n*. coppice.

CORESIE, *n*. annoyance, trouble.

CORNET PLUMS, *n*. fruit of the cornel (*Cornus mas*).

CORNETH, *v*. preserve and season, cure, i.e. pickle with "corns" or grains of salt.

COST, *n*. coast, country.

COSTMARIE, *n*. costmary, alecost (*Chrysanthemum balsamita*).

COTE, *v*. cogitate, reflect.

COTED, *v*. took note of, wrote down.

COURT, *n*. account, examination.

COUSLEPS, *n.* cowslips.
COWLASKE, *n.* diarrhoea in cattle.
CRABS, *n.* crab apples.
CRACKETH, *v.* half breaks, injures.
CRADLE, *n.* wooden three-forked instrument on which the corn is caught as it falls from the scythe.
CRAKE, *v.* brag, boast.
CRAKERS, *n.* boasters.
CREAKE, *v.* repent, "to cry creak"—to give in.
CREEKES, *n.* corners, hiding places. To seek creeks, i.e. to look for a hiding place.
CREEKES, *n.* servants.
CRESIES, *n.* cress.
CROME, *n.* stick or handle with a hook at the end.
CRONE, *v.* pick out the crones, i.e. the old ewes.
CROPPERS, *n.* best or most productive crop.
CROPPERS, *n.* persons who extract crop after crop from the land.
CROSSE, *v.* happen, result unfavourably.
CROSSEROWE, *n.* Christ-cross-row; the alphabet.
CROSSES, *n.* troubles, misfortunes.
CROTCH, *n.* curved weeding tool.
CROTCHES, *n.* crutches.
CROTCHIS, *n.* crooks, hooks.
CROWCHMAS, *n.* St. Helens' Day, 3rd May, the festival of the Invention of the Cross.
CROWE, *n.* crowbar.
CUMBERSOME, *adj.* troublesome, vexatious, oppressive.
CUNNIE, *n.* rabbit.
CURRANT, *adj.* current coin, good coin.
CURREY, *v.* gain by flattery.

DABBLITH, *v.* make wet and dirty.
DAFFADONDILLIES, *n.* daffodils.
DAIETH, *v.* names some future day for payment, i.e. buys on credit.
DAINTY, *adj.* difficult.

DALLOPS, *n.* patches of ground in the corn, or tufts of corn where dungheaps have stood in shady places.
DARE, *v.* pain, grieve.
DARTH, *n.* dearth, dearness of food.
DAW, *n.* simpleton, sluggard, chattering fool.
DAY, *n.* day-work, time-work.
DAYETH, *v.* see DAIETH.
DEAD, *adj.* flat (beer).
DEAW, *n.* dew, damp.,
DEFENDE, *v.* avoid, prevent.
DEINTILY, *adv.* dearly.
DELAIDE, *pp.* tempered, moderated.
DESCANT, *v.* comment.
DESPAIRE, DISPAIRE, *n.* injury, damage.
DET, *n.* debt.
DETANIE, *n.* dittany, pepperwort (*Dictamnus albus*).
DEW-RETTING, *v.* steeping flax by leaving it out all night on the grass.
DIALL, *n.* sun-dial.
DIDALL, *n.* triangular spade or iron scoop, for clearing ditches.
DIGEST, *v.* quiet, soothe.
DIGHT, *pp.* prepared, treated.
DIPPINGS, *n.* dripping, grease.
DISPAIRE, *v.* injure, depreciate.
DISSURIE, *n.* difficulty in passing water.
DOLE, *n.* share.
DOLES, *n.* boundary marks, also a balk or slip of unploughed ground.
DON, *pp.* done.
DOO OF, *v.* get rid of.
DOONG CRONE, *n.* crook or staff with hooked end for drawing dung.
DOTED, *v.* became foolish, was silly.
DOUSE, *n.* strumpet, prostitute.
DOUT, *n.* danger, risk, difficulty.
DOWEBAKE, *n.* dough, underbaked bread.
DRAGONS, *n.* dragon arum (*Dracunculus vulgaris*).
DREDGE, *n.* mixture of oats and barley.
DREST, *pp.* treated.
DREVE, DRIVE, *v.* drive away.

DREVILS, DRIVELL, *n.* drudge.
DRIFT, *n.* end, aim, design.
DRINKE CORN, *n.* barley.
DRIPING, *v.* dripping on, keeping wet.
DROIE, *n.* drudge, servant.
DROWSETH, *v.* becomes drowsy.
DUCK, *n.* misprint for dock.
DY, *n.* die; as close as a dy—as close as possible.

EARTHES, *n.* harrowings.
EASETH, *v.* indulges, pleases.
EDDER, *n.* flexible branches for interlacing hedging stakes.
EDISH, *n.* stubble after the corn is cut.
ELFE, *n.* tricky fellow.
ELVES, *n.* young cattle.
EMBRAID, *v.* upbraid, abuse.
EMBRINGS, *n.* Ember Days: Wednesday, Friday and Saturday after the first Sunday in Lent, the feast of Whitsuntide, the 14th September, and the 13th December.
ER, *adv.* ere, before. "Er an"—before that.
ERECTING, *pr.p.* sustaining, strengthening.
ERIE, ERY, *adj.* every.
ETCH, *n.* land still uncultivated after crop clearance, eddish.
EXCEPTIONS, *n.* differences, distinctions.
EXELTRED, *adj.* furnished with an axletree.
EXPULSED, *v.* expelled.
EY, "forgetting his eye"—neglecting his duty by staring or gaping about.

FALL, *v.* are born.
FALLETH, *v.* falls off, loses flesh.
FARE, *n.* treatment.
FARE, *v.* farrow, litter.
FAT, *adj.* as *n.* fattened beasts.
FAT, *n.* vat, vessel.
FAY, *n.* "By my fay"—by my faith, upon my word.
FEAW, *adj.* few, a few.

FEAWE, *adj.* little time, while.
FEES, *n.* pay, reward.
FEFT, *pp.* enfeoffed, endowed.
FENNIE, *adj.* mouldy.
FETCHES, *n.* tricks, stratagems.
FETHERFEW, *n.* feverfew (*Chrysanthemum parthenium*).
FIDE, *pp.* purified, cleansed.
FIE, *v.* cleanse.
FEYING, cleaning a ditch or pond.
FIEMBLE, *adj.* female.
FIERBOTE, *n.* right to take wood for burning.
FILBEARDS, *n.* filberts.
FILBELLIE, *n.* extravagance in food.
FILDES, *n.* fields.
FISGIG, *n.* worthless fellow: lightheeled wench.
FITCHIS, *n.* tares, vetches.
FITLY, *adj.* suitable, fit.
FLAP, *n.* stroke with the flail.
FLAWNES, *n.* cheesecakes.
FLEECES, *n.* frauds, impositions.
FLEERING, *v.* laughing, grinning.
FLIXE, *n.* flux.
FLOTED, *v.* skimmed off the cream.
FLOTTE, *pp.* skimmed.
FLOWER ARMOR, *n.* amaranth (*Amaranthus tricolor*).
FLOWER GENTLE, *n.* amaranth (*Amaranthus tricolor*).
FLOWER DE LUCE, *n.* yellow iris, fleur-de-lis (*Iris pseudacorus*).
FOISON, FOYZON, *n.* winter food.
FOISTY, *adj.* musty.
FOR, *prep.* in spite of, regardless of.
FOR. In numerous instances in Tusser *for* means "for fear of", "to prevent".
FORBEARER, *n.* one who refuses.
FOREHORSE, *n.* one who is always in advance with his work, the opposite to a procrastinator.
FORNIGHT, *n.* fortnight.
FOYSON, *n.* see FOISON.
FRAID, *pp.* frightened, made afraid.
FRAIE, *n.* quarrel, fray.
FRAME, *v.* make.

FRANSIE, *n.* madness, frenzy.
FRAUD, *v.* obtain by fraud.
FRAY, *n.* disturbance, trouble.
FREAT, *v.* be vexed.
FREAT, *v.* damage, decay, eat away.
FRIER, *n.* friar.
FROTH, *adj.* tender.
FROWER, *n.* a frow, an iron instrument for splitting laths.
FUMETORIE, FUMENTORIE, *n.* fumitory (*Fumaria officinalis*).
FURMENTIE, *n.* frumenty, hulled wheat boiled in milk, and seasoned with cinnamon, sugar, etc.

GADDING, *v.* going about gossiping.
GAFFE, *n.* old man, gaffer.
GAGE, *n.* pawn; "sweepeth to gage"—hurries to pledge or place in pawn.
GAP, *n.* opening, cause.
GAPING, *p.* being greedy, grasping.
GARMANDER, *n.* germander (*Teucrium chamaedrys*).
GARSONG, *n.* boy, lad.
GASING, *p.* gazing, staring.
GATE, *n.* walk, gait.
GAYLER, *n.* guardian, housekeeper.
GEANIE, *adj.* profitable, useful.
GENTILES, *n.* gentle-folk.
GENTILIE, *adv.* kindly, with proper respect.
GENTILS, *n.* gentles, maggots.
GESSE, *v.* guess, believe.
GILLET, *n.* lad.
GILOFLOWER, *n.* gillyflower, wall-flower (*Cheiranthus cheiri*).
GINNE, *n.* trap.
GINNES, *n.* means, contrivances.
GINNIE, Jenny.
GISE, *n.* fashion, way, guise.
GOD NIGHT, ? good night!
GOEF, GOFE, *n.* stack or rick.
GOELER, *adj.* more luxuriant.
GOOM,, *n.* gum.
GOSSEP, *n.* gossips, companions.
GOVE, *pp.* laid up in the barn in the straw.

GRAFFING, *n.* grafting.
GRASSEBEEFE, *n.* beef of an ox fattened upon grass.
GRATE, *n.* prison (grating).
GREASETH, *v.* bribes, enriches.
GREAT, BY, by great work, i.e. by task work, piece work.
GREEFE, *n.* trouble, worry.
GREGORIE, St. Gregory's Day, 12th March.
GROMEL, *n.* gromwell (*Lithospermum officinale*).
GROSEST, *adj.* heaviest, thickest.
GROSSUM CAPUT, blockhead.
GROTES, *n.* money (groats).
GRUCH, GRUTCH, *v.* grudge.
GRUTCHING, *n.* grumbling.
GUISE, GUYSE, *n.* habit, custom.
GUNSTONE, *n.* stone cannon-ball.
GUTTED, *pp.* taken off from the old roots.
GUTTING, *v.* cutting up, making ruts in.

HABERDEN, *n.* large salted cod.
HAFT, *v.* act like a miser, be a niggard.
HAIER, *n.* haircloth.
HAITHORNE, *n.* hawthorn.
HALLOMAS, *n.* Feast of All Saints. Hallowmas.
HALLONTIDE, All Saints' Day, 1st November.
HANDSOME, *adj.* useful, ready, handy.
HANDSOMLY, *adv.* neatly, trimly.
HARDHEAD, *adj.* hardy, brave.
HARDLIE, *adv.* with difficulty.
HARLOTS, *n.* tramps, vagrants or disreputable characters of either sex.
HARMES, *n.* injuries, pains.
HAROLDS BOOKE, *n.* Books of the College of Heralds, "visitations", county by county, giving pedigrees of armigerous families.
HART, *n.* strength, fertility, heart.
HARTED, *pp.* provided with a good heart, strengthened.
HASTINGS, *n.* early variety of peas, "soone ripe, soone rotten".

HAUNT, *v.* follow, pursue, be accustomed.
HAWME, *n.* haulm, straw.
HEARE, *n.* hair.
HEAWERS, *n.* woodcutters, hewers.
HED, *n.* head, mind.
HEW, *n.* colour, "changed hew"—have changed, become unfavourable.
HEW PROWLER, night-walker. Hugh the prowler or night-walker.
HID, *n.* care, heed.
HIER, *n.* business, duty.
HILBACK, *n.* clothing.
HINDRING, *v.* injuring, damaging.
HIR, *pron.* their.
HOBBARD DE HOY, *n.* hobbledehoy, stripling.
HOGSCOTE, *n.* pen or sty for pigs.
HOLDS, *v.* equals, gains equal.
HONE, *n.* common rubber or whetstone.
HONIE, *adj.* sweet.
HORSELOCK, *n.* shackles for horses' feet.
HOSTIS, *n.* entertainers.
HOVELL, *n.* barn, outhouse.
HOVEN, *pp.* swelled.
HOY, *v.* drag, frighten, drive away by crying, "hoy, hoy".
HULL, *n.* holly.
HULVER, *n.* holly.
HURTILBERIES, *n.* the hurtleberry or whortleberry, bilberry.
HUTCH, *n.* money chest or box.

INDIAN EIE, *n.* the pink.
INHOLDER, *n.* innkeeper.
INNED, *pp.* saved, housed.
INTREATING, *n.* treatment.
INVEST, *v.* surround.
ISE, *n.* ice.
ISOP, *n.* hyssop, (*Hyssopus officinalis*).

JACK, *n.* horse or wooden frame upon which wood is sawn.
JACK, *n.* leather tankard.
JADE, *n.* inferior horse.

JANTING, *v.* driving.
JARRING, *n.* quarrelling, scolding.
JERKE, *n.* stroke, blow.
JET, *v.* strut about, walk proudly.
JETTIE, *v.* walk or strut about.
JOBBING, *v.* pecking.
JOHN BAPTIST, feast of St. John the Baptist, 24th June.
JORNIE, *v.* go on a journey.
JUST, *adv.* neatly, trimly.

KARLE HEMPE, *n.* what was thought to be the male hemp plant.
KEIES, *n.* keys, locks.
KELL, *n.* hop-kiln.
KERVE, *v.* (carve) set out, arrange.
KEST, *v.* cast, turn.
KIFFE, *n.* kith, kindred, relations.
KINDE, *n.* nature, natural way.
KIRNELS, *n.* pips, seeds.
KNACKER, *n.* harness-maker.
KNACKES, *n.* knick-knacks, trifles.
KNOTTED, *adj.* jointed.

LAG, *v.* pilfer, steal.
LAGGED, *pp.* caught.
LAGGOOSE, *n.* Gill or Gillian Laggoose: personification of a laggard servant.
LAIE, *v.* plan, intend, purpose.
LAIE, LAY, *n.* untilled land, fallow land.
LAIER, *n.* soil, ground, kind or nature of soil.
LAIER, *n.* bed, litter.
LANGDEBIEFE, *n.* Langue de boeuf, here = borage, (*Borago officinalis*).
LARKES FOOT, *n.* larkspur (*Delphineum ajacis*).
LASH, *n.* leave in the lash—leave in the lurch.
LASH, *n.* leash by which a dog is held. "To run in the lash", to lose one's independence; hence, to fall into the snare.
LASH OUT, *v.* lavish, spend.
LASHETH, *v.* lavisheth, wastes.
LASTER, *n.* "is no laster"—will not or does not last, i.e., is soon broken.

LAUS TIBI, *n.* white narcissus (*Narcissus poeticus*).
LAVENDER COTTEN, *n.* lavender cotton (*Santolina chamaecyparissus*).
LAWE, *n.* rule; "for a lawe"—as a rule.
LAXE, *n.* looseness, diarrhoea.
LAY, *v.* plan, try.
LAY LAND, *n.* fallow land.
LEAD, *n.* cauldron, copper or kettle.
LEASE, *n.* leaze, small enclosure near the homestead.
LEAVENS, *n.* Left-overs of fermenting dough added to start the process in new dough.
LEESE, *v.* lose, miss.
LEETE, *n.* manor court.
LENT STUFFE, *n.* provisions for Lent.
LESSE, *n.* lease, term.
LET, *n.* hindrance, obstacle.
LETTED, *pp.* hindered, delayed.
LEVER, *adv.* sooner, rather.
LICOUR, *n.* water, drink.
LIDE, *v.* lay, was situated.
LIENG ALONGE, lying at a distance.
LIGHTLY, *adv.* easily, probably.
LILLIUM CUMVALIUM, *n.* lily-of-the-valley.
LINE, *n.* cord bag, carrier net.
LINNE, *n.* town of Lynn.
LITHERLY, *adj.* lazy, idle.
LIVELY SPIDE, quickly seen.
LONE, *n.* loan.
LONGING, *n.* desire, what it requires.
LONGWORT, *n.* lungwort (*Pulmonaria officinalis*).
LOP, *n.* faggot wood of a tree.
LOSELS, *n.* worthless, abandoned fellows.
LOWE, *adj.* not advanced.
LURCHED, *pp.* robbed of their food, being left in the lurch.
LURCHING, *n.* greediness.
LURKE, *v.* idle, loiter about.
LUST, *n.* desire.

MADS, *n.* blowfly maggots.
MAGGET THE PY, magpie.

MAINE WHEAT, mixed wheat.
MAINECOMBE, *n.* comb for horses' manes.
MAJEROM, *n.* marjoram.
MALE, *n.* mail-bag, portmanteau, or sack.
MAREFOLES, *n.* fillies.
MARKE, *n.* marking tool.
MARROW, *n.* mate, companion.
MARSH MEN, *n.* farmers in the fen and marshy country.
MARTINMAS, feast of St. Martin, 11th November.
MAST, *n.* fruit of the oak and beech.
MASTLIN, *n.* mixed corn.
MAWDLIN, *n.* Magdalene.
MEAKE, *n.* pease-meak, long-handled hooked blade.
MEANE, *n.* means, help.
MEANIE, *adj.* many.
MEASLING, becoming measly.
MEASURE, *v.* be moderate, be within measure.
MEAT, *n.* fed.
MEEDEFUL, *adj.* thankful.
MEEDES, *n.* meadows.
MENDBREECH, *n.* one who sits up late at night to mend his clothes.
MESTLEN, *n.* mixture of wheat and rye.
MEW, *n.* moulting cage for hawks.
MICHEL, MIHEL, MIHELL, *n.* Michaelmas. The feast of St. Michael and All Angels, 29th September.
MICHERS, *n.* lurking thieves, skulkers.
MICKLE, *adj.* great, much.
MIER, *n.* mire, filth.
MIHELMAS, Michaelmas.
MILLONS, *n.* melons.
MIND, *v.* notice, comment on.
MINION, *adj.* pleasant, agreeable, favourite.
MINNEKIN, *adj.* little.
MIRING, *v.* being stuck in bogs.
MIS, *v.* want, be without.
MISDEEME, *v.* misjudge.
MISTLE, *n.* mistletoe.
MITCH, *adj.* large.

MO, *adj.* more, others.
MOETHER, *n.* girl.
MOGWORT, *n.* mugwort (*Artemisia vulgaris*).
MOLDING, *v.* becoming musty, or mouldy.
MOME, *n.* blockhead, fool.
MOTHER, *n.* girl.
MOULSPARE, *n.* mole spear.
MOWSE, *v.* mouth, bite.
MOWTH, *v.* eat.
MULLEY, a common name for a cow in East Anglia.
MUSK MILLION, *n.* musk melon.
MYSLEN, *n.* mixed corn.

NADS, *n.* adze.
NALL, *n.* awl.
NAUGHTIE, *adj.* useless, unfit.
NAUGHTLY, *adv.* by unfair or improper means.
NAVEWES, *n.* rape, cole (*Brassica napus*).
NEATHERED, *n.* herdsman, man who attends to the cattle.
NEP, *n.* catmint (*Nepeta cataria*).
NESTLING, *v.* harbouring, supporting.
NETTIE, *adj.* natty, neat.
NIE, *adj.* near, convenient.
NIGELLA ROMANA, *n.* Love-in-a-Mist, Devil-in-a-bush (*Nigella damascena*).
NIGGERLY, *adj.* niggardly, miserly.
NOBLE, *n.* noble, a gold coin of the value of 6s. 8d.
NODDIES, NODIE, *n.* simpletons, fools.
NOE, Noah.
NOIANCE, *n.* injury, trouble.
NOIE, *v.* are injurious, noxious.
NOWLES, *n.* hillocks, little mounds, knolls.
NOY, *v.* to harm injure.
NOYER, *n.* one that hurts or injures.
NURTETH, *v.* butts or pushes with the horns.
NURTURE, *n.* training.

OFCORNE, *n.* offal or waste corn.
OPE GAP, *n.* one who opens up a gap.
OPEN, *v.* bark, open his mouth.
OPTE, *v.* opened.
OR AND, before.
ORENGIS, *n.* oranges.
OTHING, one thing.
OUT, *adj.* outdoor, open air.
OVERCOME, *v.* manage, keep up with.
OVERLY, *adv.* all over.
OVER REACHING, cheating, deceiving.
OVERTHWART, *prep.* across.
OX BOWES, *n.* the bows of wood which go round the neck of an ox.

PAD, *n.* padlock.
PAGGLES, *n.* cowslips.
PAINFULL, *adj.* painstaking, careful.
PAINFULL, *adj.* full of trouble, requiring care.
PANEL, *n.* a pannier.
PARE, *v.* injure, damage, impair.
PARED, *pp.* cleaned and cleared of all superfluous roots.
PAS, *v.* care.
PASK, *n.* Easter.
PASSETH, *v.* thinks, reflects.
PATCH, *n.* clown, dolt, i.e. farm labourer.
PATCHES, *n.* places where the shearer has cut the skin of the sheep, wounds.
PATES, *n.* persons.
PATIENCE, *n.* monk's rhubarb (*Rumex patientia*).
PAUNCIES, *n.* pansies, heartsease.
PAY, *v.* pay home—give a strong, sharp blow.
PEAKE, *v.* to look thin or sickly.
PEASEBOLT, *n.* pease-straw.
PEASEETCH, *n.* pease-stubble.
PEASON, *n.* pease.
PECK, *n.* a peck measure.
PED, *n.* a pannier, a large capacious basket.
PEELER, *n.* an impoverisher.
PELFE, *n.* apparatus, implements.
PENERIALL, *n.* pennyroyal.
PENIE, *n.* penny, money.

PERCELEY, *n.* parsley.
PERCER, *n.* a piercer, gimlet.
PERSENEPS, *n.* parsnips.
PESTER, *v.* overcrowd with stock.
PESTRING, *v.* being in the way or troublesome.
PHILIP AND JACOB, The feasts of Saints Philip and James, 1st May.
PHRAIES, *n.* phrase, language.
PICKLE, *n.* condition, state.
PIDDLING, *v.* going about pretending to work but doing little or nothing.
PIE, *n.* magpie.
PIKE, *n.* a pitchfork, put into sheaves.
PILCH, *v.* pilfer.
PILFERIE, *n.* theft, fraud.
PINCHING, *n.* economy.
PINWOOD, *n.* wood suitable for pins or pegs.
PIONEES, *n.* peonies.
PISMIER, *n.* ant.
PITCH AND PAY, Pay ready money.
PLAGARDS, *n.* commissions, instruments.
PLANKED, *pp.* boarded.
PLANTINE, *n.* great plantain (*Plantago major*).
PLASH, *v.* to lower and narrow a broad-spread hedge by partially cutting off the branches and entwining them with those left behind.
PLOT, *n.* plan, rule.
PLOUGH MONDAY, The first monday after Epiphany, on which the ploughing season began in East Anglia.
PLOUGHSTAFF, *n.* a small spade for clearing the mould-board and coulter of a plough.
PLOWMEAT, *n.* food made of cereals.
PLUMP, *v.* thrown in.
POKE, *n.* a bag sack, "buy a pig in a poke"—to buy without seeing what one is buying.
POLING, *n.* supporting with poles.
POLLARD, *n.* beardless, awnless wheat.
POLLENGER, *n.* pollard tree.

POMPIONS, *n.* pumpkins.
PORET, *n.* a scallion: a leek or small onion.
POSIE, *n.* a poetical inscription or motto.
POTTAGE, *n.* pottage, soup.
POTTLE, *n.* a measure of two quarts.
POUCHETH, *v.* pockets.
POUND, *v.* fight, beat.
POWLINGES, *n.* the branches or shoots of pollard trees.
PRACTISIE, *n.* conduct, practices.
PRAIES, *n.* praise.
PRAY, *n.* prey, booty, plunder.
PREST, *adj.* ready.
PREST, *adj.* neat, tidy.
PREST, *pp.* pressed.
PREVENTING, *p.* anticipating.
PRICKETH, *v.* makes proud or puffs up.
PRICKING, *v.* embroidering, doing fancy work.
PRIDE, *n.* excessive richness.
PRIE, *n.* privet.
PRIM, *n.* another name for the privet.
PRIME, *n.* the time of the new moon, as change is the time of the full moon.
PRIME GRASS, *n.* earliest grass.
PRIVIE, *adj.* aware, acquainted.
PRIVIE, *n.* privet.
PROCURETH, *v.* contrives, brings about.
PROMOOTERS, *n.* informers, professional accusers.
PROVISION, *n.* foresight.
PULLEIN, PULLEN, *n.* poultry, fowls.
PULTER, *n.* fowl keeper or breeder.
PURKEY WHEAT, turkey wheat, maize.
PUTTOCKS, *n.* kites, hawks, voracious fellows.

QUAILE, *v.* fail.
QUAMIER, *n.* quagmire, bog.
QUEENES GILLEFLOWERS, *n.* dame's violets (*Hesperis matronalis*)
QUEERE, *n.* choir.
QUIETER, *adv.* more easily.
QUIGHT, *adv.* completely, entirely.

QUITE, *v*. requite, repay.

RABETSTOCK, *n*. a rabbet-plane, a joiner's tool for cutting rabbets.
RABLE, *n*. crowd, number.
RAGE, *adj*. wild, dissipated.
RANKER, *n*. ill-feeling, quarrelling.
RASKABILIA, *n*. rascals.
RATLING, *n*. the rattle, rattling noise in the throat.
RAWING, *n*. rowen, second growth of grass, aftermath.
REAME, *n*. kingdom, country.
REASTIE, *adv*. rancid.
REDELE, *n*. riddle.
REEDED, *pp*. thatched with reeds.
REEDING, *n*. reading, study.
REEKE, *v*. smoke.
REFRAINE, *v*. stop, prevent.
REHERSED, *pp*. mentioned, named.
REISONS, *n*. red currants and blackcurrants.
RELENT, *v*. become soft.
RENDRIT, *v*. render it, *i.e.* return, requite it.
REPT, *pp*. reaped, gained.
RESPE, RESPIES, *n*. raspberries.
RETCHELES, *adj*. reckless, careless.
REW, *n*. rue (*Ruta graveolens*).
RIFLE, *n*. bent stick on the butt of a scythe-handle for laying the corn in rows.
RIGGING, *p*. making free with, knocking about.
RIGS, *v*. makes free with.
RINGLE, *v*. ring, put rings through the snouts.
RIPING, ripening.
RIKES, *n*. ricks.
RISE, *n*. rice.
RISHES, *n*. rushes.
RIVET, *n*. bearded wheat.
RODE, *n*. harbour.
ROINISH, *adj*. mean, rough, coarse.
ROISER LIKE, blustering.
ROKAT, *n*. garden rocket (*Eruca sativa*).
ROONG, *pp*. have rings put through their noses to prevent them from tearing the ground.
ROPERIPE, *n*. one old enough to be flogged.
ROTTENLY, *adj*. rich, crumbly.
ROULE, *n*. a rule, measure.
ROULES, *v*. rolls in, brings in.
ROWE, *n*. row; "a rowe"—in a row.
ROWEN, aftermath of mown meadows.
RUBSTONE, *n*. a sandstone for a scythe.
RUFFEN, *n*. ruffian, scoundrel.
RUNCIVALL PEAS, *n*. marrow-fat peas.
RUNNAGATE, *n*. runaway.
RYDGIS, *n*. ridges.

SAD, *adj*. disappointed, vexed.
SAILE, *n*. to bear low sail, i.e. to live humbly or economically.
SALLETS, *n*. salads.
SALLOW, *n*. low-growing kinds of willow.
SAVER, *n*. scent, inkling.
SAVER, *n*. a person to look after and see that things are not wasted.
SAVERIE, *n*. savory (*Satureja montana*).
SAVERLIE, *adj*. frugal, gained by saving.
SCAMBLE, *v*. scramble for.
SCANT, *adv*. scarcely.
SCANTED, *adj*. limited, stinted, grudged.
SCANTETH, *v*. is economical.
SCAPE, *v*. escape, get off.
SCOTCH, *v*. cut, hew
SCRALL, *v*. crawl.
SCRAULING, *p*. crawling.
SEA HOLIE, *n*. sea holly (*Eryngium maritimum*).
SEALED, *adj*. certified, stamped.
SEAME, *n*. pack-horse load of corn, eight bushels.
SECRESIE, *n*. secrets, private concerns.
SEDGE COLLARS, *n*. collars made of sedge or reeds.
SEEDE, *v*. obtain seed from.
SEEDE CAKE, cake flavoured with caraway seed eaten at the festival which marked the end of sowing.

SEEITH, *v.* boil.
SEELIE, *adj.* silly, simple.
SEENE, *adj.* practised, experienced.
SEETH, *v.* boil.
SEEVE, *n.* sieve, sifter.
SEGGONS, *n.* poor labourers.
SEITHES, *n.* chives.
SELL, *n.* cell, abbey.
SERVITURE, *n.* servant, attendant.
SET, *n.* the young shoots.
SETTETH, *v.* risks.
SEVER, *v.* separate, sort.
SEVERALL, *n.* inclosed land, divided into fields by fences.
SEWE, *v.* drain.
SHACK TIME, *n.* when "shack"—grain fallen from the ear—lies about in the stubble for farm animals and poultry to pick up.
SHARE, *v.* shear.
SHARING, *pr. p.* shearing.
SHAVE, *n.* spokeshave.
SHED, *v.* lose the grains of corn.
SHEEPEBITER, *n.* thief.
SHENT, *pp.* ruined, disgraced.
SHERE, *n.* shire, county.
SHIFT, *v.* manage, fare.
SHIFTING, *pr. p.* changing, often removing.
SHIFTING, *n.* trickery.
SHOD, *pp.* tired.
SHOLVE, *n.* shovel.
SHOT, *n.* expense, reckoning.
SHOWRETH OUT, *v.* is showery, rainy weather.
SHREAW, *n.* thief, rascal.
SHRED PIES, *n.* mince pies.
SHROFTIDE, *n.* Shrovetide, Quinquagesima Sunday, Shrove Monday and Shrove Tuesday.
SHROVING, *n.* making merry at Shrovetide, before the rigours of Lent.
SHUT, *v.* shoot, throw.
SIETH, *n.* scythe.
SISZERS, *n.* scissors.
SKAVEL, *n.* spade, having its sides slightly turned up, used in draining and cleaning narrow ditches.
SKEP, *n.* basket made of rushes or straw.
SKILL, *n.* plan, design.
SKILLESSE, *adj.* simple, homely.
SKIRRETS, *n.* the root vegetable *Sium sisarum*.
SKREENE, *n.* fire-screen.
SKREINE, *n.* sieve, screen.
SKUPPAT, *n.* spade used in draining and making narrow ditches.
SKUTTLE, *n.* screen for cleaning corn.
SLAB, *n.* outside cut of sawn timber.
SLABBERED, *pp.* dirtied, beslobbered.
SLAKE, *v.* to slacken.
SLAPSAUCE, *n.* glutton.
SLEA, *v.* slay, kill.
SLEPT, *pp.* slipt, forgotten, omitted.
SLIVERS, *n.* pieces of split wood, chips.
SLUGGING, *n.* lying late in bed.
SMACK, *n.* pleasant repast.
SMALACH, *n.* celery.
SMALLNUTS, SMALNUTS, *n.* hazel (*Corylus avellana*).
SNORTING, *adj.* snoring.
SNUDGETH, *n.* is economical or saving, or, works quietly or snugly.
SOCKLE, *v.* suckle, provide with milk.
SOD, *pp.* boiled.
SOLES, *n.* collars of wood, put round the neck of cattle to confine them to the post.
SOLLEN, *adj.* sullen, sulky.
SOLLER, *n.* garret, loft, or upper room.
SOOTH, *v.* to flatter.
SOPS IN WINE, *n.* clove-pink (*Dianthus caryophyllus*).
SOST, *pp.* dirty, foul.
SOUSE, *n.* pigs' feet and ears pickled.
SOUTAGE, *n.* bagging for hops or coarse cloth.
SOWCE, *v.* steep in brine, pickle.
SOWER, *adj.* sour.
SPARS, *n.* rafters.
SPEEDFULL, *adj.* useful, profitable.
SPEEDING, *n.* progress, success.
SPEERED, *pp.* sprouted, germinated.

SPENT, *pp*. used, consumed.
SPERAGE, *n*. asparagus.
SPIALS, *n*. spies.
SPIDE, *v*. beheld, saw.
SPIGHT, *n*. grief.
SPIGHT, *v*. spite, be unpropitious.
SPIL, *v*. spoil, ruin.
SPILLED, SPILT, *pp*. ruined, spoilt.
SPRING, *n*. shoots, suckers, young growth.
SPURLINGS, *n*. smelts.
SQUIER, *n*. squire, gentleman.
STADDLES, STADLES, *n*. young trees left standing when a wood is cleared.
STADLE, *v*. leave at certain distances a sufficient number of young trees to replenish a wood.
STAID, *v*. kept, detained.
STAIE, *n*. means of support.
STAIE, *v*. prevent, stop.
STAIED, *adj*. steady, staid.
STALFED, *adj*. stall-fattened.
STAMP, *v*. bruise, pound.
STANDS UPON, are incumbent on.
STAR OF JERUSALEM, *n*. ? vegetable oyster, salsify (*Tragopogon porrifolius*).
STAY, *n*. rest, quiet.
STEADE, *n*. "in steade"—to advantage.
STEDE, *v*. suffice, profit.
STEELIE, *adj*. hard, firm.
STEEPE, *adj*. "a steep", steeply.
STERVELINGS, *n*. half-starved animals.
STICK, *v*. "to stick boards"—to arrange them neatly one upon another with sticks between.
STILL, *v*. quiet, stop from growing.
STILL, *v*. distil.
STINTED, *pp*. appointed, settled.
STIRRE, *v*. move quickly, bestir.
STOCKE GILLEFLOWERS, *n*. stocks (*Matthiola incana*).
STOCKS,, *n*. young trees.
STOUTNESSE, *n*. force.
STOVER, *n*. winter food for cattle.
STREIGHT WAIES, *adv*. at once.
STRIKE, *n*. bushel measure.

STRIPE, *n*. beating.
STROIENG, *n*. destruction, injury.
STROKEN, *pp*. kindly treated.
STROWING, *adj*. for strewing.
STROYAL, *n*. waste all, wasteful.
STUB, *v*. grub up.
STUB, *n*. stump; "buie at the stub"—buy on the ground.
STUD, *n*. uprights in a lath and plaster wall.
STUR, *v*. move about, exert.
STURS, *n*. distubances, commotions.
SUBSTANCIALLIE, *adv*. in reality, truly.
SUBTILTIE, *n*. cunning, artfulness, deceit.
SUCKER, *n*. assistance, help, succour.
SUCKERIE, SUCKERY, *n*. succory, chicory (*Cichorium intypus*).
SUDGERNE, *v*. sojourn.
SUERTY, *n*. being security or surety.
SUITE, *n*. description, kind.
SWAGE, *v*. assuage.
SWATCHES, *n*. rows or ranks of barley, etc.
SWATHES, *n*. line of grass or corn cut and thrown together by the scythe in mowing.
SWEATE, *n*. sweating, *i.e.* feel the effects of the heat.
SWEETE JOHNS, *n*. species of pink (*Dianthus barbatus*).
SWERVE, *n*. fail, depart.
SWILL, *n*. hog's wash.
SWIM, *v*. abound, overflow.
SWINGE, *v*. cut down with the long swinging scythe used for that purpose.

TACK, TACKE, *n*. substance. A tough piece of meat is said to have plenty of *tack* in it.
TAILE, *n*. back.
TAINT WORMS, *n*. worms or grubs believed to "taint" or infect livestock.
TALE, *n*. tally, reckoning.
TALLIE, *n*. score, bill, charge.

TALLWOOD, *n*. wood cut for billets.
TAMPER, *n*. condition.
TAMPRING, *v*. tempering, mixing.
TANE, *pp*. taken.
TAPPLE UP TAILE, turn topsy-turvy and die.
TARIE, *v*. delay, keep back.
TEDDER, *n*. tether, "live within one's tether"—within the limits of one's ability or scope.
TEEMES, *n*. teams.
TELL, *v*. count.
TEMMES LOFE, *n*. ? Thames loaf, made of fine wheat and rye.
THACKE, *n*. thatch.
THEE, *v*. thrive, prosper.
THEEVERIE, *n*. dishonesty.
THENCREASE, "for the encrease"—the increase, gain.
THEND, the end.
THETCH, *n*. thatch.
THICKER, *adv*. more frequently.
THIES, *n*. thighs, limbs.
THILLER, *n*. shaft-horse, last horse in a team.
THOES, *pr*. those.
THON, the one.
THOROW, *v*. pass through.
THOTHER, the other.
THRESH, *v*. whip, thrash.
THRESHER, *n*. duster of furniture with a goose wing.
THRIFT, *n*. fortune, success, prosperity.
THRY-FALLOW, *v*. plough fallow land three times.
TIBURNE STRETCH, Tyburn stretch, a hanging.
TIDE, *pp*. tied, fastened.
TIDIE, *adj*. neat, proper, or in season.
TIETH, *n* tithe.
TILMAN, *n*. farm labourer, ploughman.
TILTH, *n*. tillage, cultivation.
TILTH, *n*. ground tilled.
TILTURE, *n*. tillage, cultivation.
TINE, *n*. wild vetch or tare.
TITS, *n*. horses.

TITTERS, *n*. noxious weed amongst corn.
TODE WITH AN R, *n*. a turd.
TOESED, *pp*. pulled, pinched.
TOIENG, *p*. playing, amusing.
TOIES, *n*. amusements, occupations.
TOLLETH, *v*. takes toll.
TON . . . TOTHER, the one, the other.
TONE, the one.
TOOTETH, *v*. looks or strives anxiously.
TOST, *v*. agitated, harassed.
TOUCH, *n*. faith, honour; "to keep touch," perform a promise.
TRAIE, *n*. mason's hod.
TRAINE, *v*. draw.
TRANSPOSE, *v*. arrange, dispose of.
TRAVERSE, *v*. start upon, proceed upon.
TREENE, *adj*. wooden.
TRICK, TRICKLY, *adj*. neat, clean, tidy.
TRICKETH, *v*. dresses up, furnishes.
TRIM, *v*. repair.
TRIM, *adv*. quickly, at once, easily.
TRIMLIE, *adv*. neatly, cleanly.
TRINKETS, *n*. porringers.
TRIVE, *v*. (for contrive) try, attempt.
TROFFE, *n*. trough.
TROPE, *n*. phrase.
TROTH, *n*. truth.
TROWLETH, *v*. helps on, moves towards.
TRUDGE, *v*. go, be spent.
TRUDGETH, *v*. labours, journeys far.
TRULL, *n*. girl, lass.
TULLIE, Cicero.
TUMB, *n*. the tomb, grave.
TURNEBROCH, *n*. turnspit, boy who turns the broach or spit.
TURN UP, *v*. deck, ornament.
TWELFTIDE, *n*. Twelfth Day, *i.e.* January 6th, twelve days after Christmas.
TWIFALLOW, *v*. till twice, plough twice.
TWIGGERS, *n*. prolific breeders.
TWIGGING, *n*. fast breeding.
TWINLINGS, *n*. twins.

TWINNING, *n.* the bearing of twins.
TWITCHER, *n.* instrument used for clinching the pig rings.
TWITCHIS, *n.* sharp pains, pinches.

UNDESKANTED, *pp.* untalked of.
UNFAINEDLIE, *adv.* unfeignedly, in truth.
UNLUSTIE, *adj.* poor.
UNMEETE, *adj.* unfit.
UNSAVERIE, *adj.* wasteful, ruinous.
UNSHAKEN, *adj.* perfect, in good order, free from shakes.
UNSPILT, *pp.* not wasted.
UNTANGLED, *pp.* freed from the hop vines.
UNTHRIFT, *n.* a prodigal, spendthrift.
USHER, *n.* doorkeeper.

VAINE, *n.* liking, fancy.
VAINFULL, *adj.* vain, fickle.
VANCE, *v.* advance.
VEGETIVE, *adj.* belonging to the plant.
VENT, *n.* sale, disposal.
VENTER, *v.* risk, venture.
VERGIS, *n.* verjuice, acid juice of crab apples.
VERIE, *adj.* true, real.
VERLETS, *n.* rascals, scoundrels.
VERMIN, *v.* destroy vermin.
VEW, *n.* view, sight.
VEWE, *v.* view, examine.
VICE, *n.* buffoon.
VITLETH, *v.* eats, dines.
VOYD, *v.* avoid.

WADLING, *n.* wattling, wattled fence.
WAG, *n.* messenger.
WAID, *pp.* considered, reflected on.
WAIETH, WAITH, *v.* considers, reflects.
WAIGHT, *v.* watch, wait about.
WAIGHT, *v.* wait at table.
WAINE, *v.* fetch, bring.
WAKE DAY, *n.* day of the village wake or festival, on the feast of the patron saint of the church.
WALKE, *n.* pasturing.
WAND, *v.* inclose with poles.
WANTETH, *v.* is in want, is without.
WANTEY, *n.* rope or leather girdle, by which packs were fastened to a pack-saddle.
WANTON, *n.* merry girl.
WARELY, *adv.* carefully, warily.
WARES, *n.* productions.
WARRENER, *n.* keeper of a warren.
WART, *v.* wert, wast.
WASTER, *n.* wasteful thing.
WATER-RETTING, steeping flax in water to separate the fibres.
WAYEST, *v.* considerest.
WEATHER, *v.* dry in the open air.
WEBSTER, *n.* weaver.
WEDEHOKE, *n.* hook for cutting away weeds.
WEELES, *n.* traps for fish made of osiers.
WEENE, *v.* think.
WEFTE, *n.* smoky flavour.
WELL A FINE, to a good end or purpose.
WENNEL, *n.* calf just weaned.
WHEAT PLUMS, *n.* variety of plum.
WHEELE LADDER, *n.* probably the frame on the side of a cart to support the hay or corn when the load is to be increased.
WHELPE, *n.* child.
WHEREAS, *adv.* wherever.
WHIGHT, *adj.* white.
WHINNES, *n.* whin, furze.
WHIPSTOCK, *n.* handle of a whip.
WHIST, *v.* be silent, be hushed.
WHITCH, which(kind).
WHIT LEATHER, *n.* leather dressed with alum and salt, to retain its white or pale colour.
WHITEMEAT, WHITMEAT, *n.* food prepared from milk.
WICKET, *n.* small doorway i.e. (woman's) mouth.
WILFULL, *adj.* ready, hasty.
WIMBLE, *n.* auger.
WINE, *v.* win, make to please.
WIT, *n.* sense, good judgement.
WITHER, *v.* dry.

WONNE, *pp.* managed, made up.
WOOD, *adj.* mad.
WOODROFE, *n.* woodruff (*Asperula odorata*).
WOODSERE, *n.* season in which a tree will decay or die if its wood is cut.
WOORSER, WORSER, *adv.* worse, a double comparative.
WOORTH, *n.* "take in woorth"—take for what I am worth.
WOT, WOTE, *v.* know.
WRALL, *v.* squall.
WRAUGHT, *pp.* supplied, furnished.
WRAULING, *n.* bawling, squalling.

WRECKE, *v.* wreak, vent.
WREST, *v.* turn, force away.
WREST, *v.* steal away, plunder.
WRESTING, *n.* struggling for, fighting for.
WRIGHT, *v.* write.
WRINGER, *n.* extortioner.
WRITE, *v.* mark, write the name on.
WUD, *n.* wood.

YARN, *v.* earn.
YEANE, *v.* bring forth young.
YEERLIE, *adv.* quickly.
YERKE, *v.* kick.

INDEX

accounts, keeping, 20
acorns (*see also* mast): harmful to cows, 36, 224; sowing, 39, 45, 233, 245; diminishing supply of, 233
ages: of man, 130, 131; of woman, 184
alder, 71, 127, 273
ant-hills, 239, 246
ants, 246
arbours, 78, 248
ash, 98, 275, 308
autumn: winds of, 25; compared with manhood, 59

baking, 167
barley: preparation of ground, 38, 41, 249; sowing, 39, 79, 86, 93, 249, 265, 267; threshing, 40, 49, 85; soil for, 42, 229; improverishes land, 43; and rotation of crops, 44, 102; malting, 47, 49, 235; manure, 80, 124, 126; clearing out mow, 81; rolling, 87, 94; weeds among, 93, 265–6; harrowing, 93, 94, 267; harvesting, 118, 123, 304; tithing, 123; storing, 124, 305; sprat or battledore-, 232; cleaning, 235; impediments to cultivation of, 266
barn tools, 31, 217
barns and outhouses, 111, 112, 293–4
basil, 90, 103, 108, 260, 290
beans: sowing, 34, 83; setting, 50, 251; garden beans, 68, 116, 299; harrowing, 83, 251; gathering, 114, 116, 299; thrive with weeding, 222; weed-destructive, 226; soil suitable for, 233, 250; influence of moon on, 251; horse fodder, 252
Bedfordshire, 296
bed times, 177, 319
beech, 308
beef, *see* meat
bee-keeping: driving of hives, 28, 35; preserving of bees, 35; winter feeding, 54, 58, 242–3; swarming, 103, 106, 287; suffocation of bees, 223; short-sight, 223
bequests, 22
Berkshire, 253
binding with hook and line, 225
birch, 308
birds: scaring, 27, 34, 87; destruction of crows, etc., 95, 271; plunder corn, 222, 283; attack graftings, 265
bishop, cursing the, 280
blackthorn, 275
blood-letting, 53, 57, 242, 318–19

borrowing, 19, 27, 111, 292
bows, 42, 227
brakes: uses of, 28, 37, 225; fuel, 46; mowing, 112, 121, 301; advantage of short-cut, 295
bramble: for hedging, 40; sowing, 45, 234; mowing, 112, 295; raising, 234; uses of, 295
brank, *see* buckwheat
bread, 222
breakfast, 164
brewing, 40, 167, 305
briony, 248
broom, 70
buckwheat (brank), 114; enriches soil, 43, 231; sowing, 102, 105, 285; gathering in, 116; uses of, 285; cattle fodder, 299; to be stored dry, 300
buying and selling, 119; timber and bark, 98, 273–4; three kinds of, 126, 307; at the fair, 127

Cambridge: Trinity Hall, xiii–xiv, 204, 209, 315, 316; lawlessness in, 136–8
Cambridgeshire, 97, 254, 272, 296
candle making, 174
carts: strong and light, 31, 217; implements, 32; timber suitable for, 98, 274; maintenance of, 111, 118, 292; cartshed, 111, 292; dredge wheels, 246, 306
cats, 50, 164, 168
cattle: gelding, 27, 35, 70, 77, 247; best cattle, most profit, 30, 216; strong and light, 31, 216; take harm from acorns, 36, 224; burning of dead cattle, 45; dry litter for, 48; calf-butting, 48, 53; feeding, 53, 54; winter care of, 53; feeding, 55, 56, 57, 96, 242, 245, 252, 257, 288; housing, 55, 240; medicine, 69, 75; calves, 69–70, 76–7, 103, 107, 247; for meat, 73, 243; browsing, 74; weaning, 76, 247; milchcow, 78, 119, 126, 248, 308; treatment of, 80, 84; provender for plough oxen, 84; on pasture land, 85; proportion of ewes to cows, 102, 104; separation of bulls from cows and heifers, 103; watering, 107; keeping oxen from cows, 108, 290; leave milking, 119; fattening, 127; evening tending, 174; collars, 221; turnip-fed, 240; slaughtering, 247; well-fed, 252; hay-fed, 273; poor man's cow, 276
chaff, 40, 49, 217, 252
cheese: making, 100–1; selling, 119; Suffolk and Essex, 278, 309; mouldy, 279–80; ewes'-milk, 282
Cheshire, 313

337

children: picking seed wheat, 38; bird scaring, 42; cattle foddering, 56; to gather stones, 107, 288; corn picking, 125, 306; housewife and, 160; mother and, 180; teaching, 182–3; use of bow by, 227; minding the cow, 276
chimney sweeping, 48, 51
Christmas, 24, 62, 63, 65, 66–7, 319
cockcrow, housewifery and, 161–2
common land: lack of profit, 99; livestock on, 99; cultivation of, 232; poor man's cow on, 248; of questionable benefit to the poor, 276; abuse of, 276; enclosure of land, 277; rich and poor and, 312
cookery: brawn, 54; tainted meat, 163; instructions to housewife, 168; Shrovetide, 177; at sheep shearing, 178; at end of wheat sowing, 178
corn: sowing, 27, 221; weight of sheaf, 31, 217; value of old corn, 40; in rotation of crops, 44; old better than new, 46; weeding, 105, 284; protection of, 111; reaping, 117; harvest, 118, 122; mow-burnt, 123, 304; expenditure of profit from, 128; seed, 221; quality of, 231; assessing yield, 238; bread-corn, 258; Hillman on growing of, 283; storage, 293; affected by moisture, 293; dallops, 298, 304; allowed to stand too long, 302; threshing, 305; damaged by swine, 310
cows, see cattle
crab apple, 27, 35, 40, 245
crab tree, 96, 98, 275
crows: scaring, 27, 34, 86, 87, 220; killing, 42, 95, 271; edible, 271

dairy (see also cheese): good and ill housewifery, 99; sale of produce, 119; instructions to housewife, 168; milk of common-grazed cows, 276; Suffolk and Essex, 278; butter, 278, 280; mistress and, 278–9; cream, 279; churning days, 289
dairymaid, a lesson for, 100–1
deer, 68, 74, 245
Devonshire, 279, 304
dicing, 20
diet: farmer's daily, 23–4, 318; and health, 179
digging, 51, 71, 113
dinner, instructions to housewife, 170–2
disorders in livestock: measling, 35, 45, 224; in fat hogs, 45; bots, 57, 298, 304; ailing cattle, 75; mads, 102; gripes, 242; laxative, 104, 282
distilling, 108, 289–90
ditches, draining, 103, 106, 112, 295–6, 310
ditching, 27
dogs: thieves', 29, 37; and hogs, 48, 88, 96, 277; folly of keeping beyond one's means, 50, 236–7; and lambs, 69; and rabbits, 81, 250; and sheep, 86, 92; scraps for, 176; to chase away peacocks, 253; mole catchers, 255; kennel, 293; to control swine, 310
dovecote, 68, 74, 244
doves (see also pigeon): feeding, 68, 74, 81; dung, 68, 74, 244; peas to be protected from, 118, 124
drainage: water furrough, 34, 38; of ditches, 103, 106; of wet ground, 257, 266; and land intended for hay, 287
dredge (bullimong), 45, 49, 233, 235
drinks, verjuice and perie, 40
drought, 262
dung, see manure

eddish, 41, 226
elm: lopping, 75; use of timber, 98, 275; Dovercourt, 219; inimical to hop-growing, 263, 264
Ely, Isle of, 249
Ember Days, 24, 215
enclosure of land, 135, 277, 309, 311
Essex: Tusser's native county, xi, 202, 315; Rivenhall, xi, xv, 202, 315; Fairstead, xvi, 208, 316; Brantham, 42, 228; enclosure of land, 135; Shrovetide, 177, 319; animal husbandry, 243; dairy farming, 278; Sturbridge Fair, 308
Eton College, xiii, 203, 315

fallowing, see ploughing
farms, 30–1, 216
feathers, 49, 173, 236
fencing, 262; care of, 29, 37, 38, 40; of pasture, 30; wood for, 75, 246; stubbing up bushes by, 76; corn, meadow and pasture, 92; before felling, 98, 274; hopyard, 113; edder, 245; on unenclosed land, 251–2; willow for, 263
fens, 97, 103
fern, to destroy, 284
fish: farmer's diet, 23, 24, 318; September husbandry, 29; salt and dried, 56, 125, 241; thieves, 82; salt cod (haberden), 119; provision for Lent, 125; burnt to the stone, 307; fish-days, 318
fishing: ponds, 82; trade, 258–9
flax, 102, 106, 114, 116, 299
fleas: remedy against, 116; to rid a bed of, 300
Flemings, 45, 234
flour, to be made weekly, 48, 51
flowers for window and garden, 90–1
fodder, cattle: stover, 48, 49, 236; chaff, 49; in December, 56; vetch, 57; provision of, 98; hay, 115, 273, 298; feathers in, 236; browse, 75, 245; turnips, 245

338

fodder, horse: care of, 57; hay, 115, 273, 298; feathers in, 236; beans, 252
fodder, pig, 233
fodder, sheep, 115, 298
foddering, 55, 240
folding, 99, 104, 277
food: at Christmas, 65; veal, pig and lamb, 73, 243; roast meat, 178, 243; broth, 243; Hillman on neglect of vegetables and herbs, 267–8, 271
football (campe), 54, 58
foxes, 81, 86
friendship, 187
fruit: gathering, 27, 28, 35, 40, 223; setting, 36, 68; strawberries, 54, 57; to be set or removed, 72; grafting, 86; perareplums, 243
fuel, 289; to economise on, 46; logs, 53, 119; splitting wood for, 55; for drying fish, 56; lop, 69; purchasing, 74; ill husbandry of, 74; felling, 75; fetching home, 103, 107; protecting from theft, 103, 281; time for city dwellers to buy, 108; bringing in and stacking, 125, 306; for malting, 170; light-firing, 235, 301, 303; to check wastage of, 244; colliers, 289
furze, 71, 112

garden: trenching, 51, 238; January work in, 68, 74; ploughing, 70, 77; best aspect, 87; comprehensive instructions on gardening, 94–5, 270; Hillman on, 224, 242, 268–9; peas, 253; peacocks in, 253; produce neglected, 267–8; sowing times, 269
garlic, 47, 50, 237
geese, for Harvest Home, 178
gelding: bulls, 27, 35; rams, 27, 35, 77; calves, 70, 77, 247; lambs, 70, 87, 247; pigs, 70, 77, 248; colts and fillies, 70, 77, 248; best time for, 223; reason for delaying, 243; spading, 248
gillyflower, 29, 57, 90, 261
gleaners, 18, 124, 246, 277, 305
grafting (graffing), 86, 93, 264
grain storage, 48
grass: to be cleared of rubbish, 69; calves out to, 103, 107; mowing, 112, 298; haymaking, 115; breaking up grassland, 249; oats and, 249; new sorts of, 275; laying of headlands for, 276; and taste of butter and cheese, 278; roots to be burnt, 297
grist, 114, 116
guests, slovenly, 187–8

harrowing: September, 34; February, 80, 83, 84; molehills, 85; barley, 93; corn, 94; buckwheat, 105

harvest, 117–18; tools, 33; hay, 115; drum and horn at, 117; work after, 118–20, 124–8; day-work and piece-work, 122, 302; good points for, 122; time of good cheer, 124; labourers, 124; expenditure of profit from, 128; Lord of, 303; beer for, 305; rewards after, 305
harvest home, 178, 319
hawking, 81, 119, 126, 312
hawthorn, 45, 72, 233, 252–3
hay: for draught oxen, 80, 252; meadows, 86; mowing, 111; haymaking, 114, 291, 297–8; harvesting, 115; buying, 249; as fodder, 273, 298; mow-burn, 273, 291, 298; haystacks, 293
hazel, 98, 275
hedging: privet and box, 29; hedge breakers, 37; quickset, 39, 45, 73, 80, 84, 225, 244, 252–3; reinforcement, 40, 225; in January, 69, 73, 74; plashing, 80; weeding, 106, 286–7; planting, 225; space cleared of bushes, 246; over burrows, 250; hedging enclosures out of the common field, 254; hedge greens, 298, 299; against rabbits, 310
hemlock, 70, 106
hemp: September husbandry, 28, 35–6; effect on nettles, 84, 254; sowing, 102, 106, 254; culling, 114, 116; uses of, 116, 299; retting, 224; linen from, 254; fimble and carle, 286, 299
herbs, 48; setting, 29, 81; for the kitchen, 88–9; for salads and sauce, 89; to boil or butter, 89–90; for strewing, 90; for windows and pots, 90–1; distilling, 91, 108, 289; for physic garden, 91, 179; remedial, 259–62; sowing, 269
Hertfordshire, 240
hewers, 75, 98, 245, 274
highways, repairing, 119
hogs, see pigs
holly (hulver), 40, 98, 274
hook and line for binding, 225
hop-poles, 75, 96, 98, 273, 308
hops: setting, 81, 92; layout of hop-hills, 92–3, 263; growing conditions, 86, 93, 248; tools, 86, 93; growing, 102, 106, 264; weeds inimical to, 106; properties of, 113, 297; husbandry, 119–20, 127–8, 308; attention to roots, 263; fencing, 263; twine round hop-poles, 264; early use of, 296; best soil for, 296; colour of, 308; drying, 309; packing, 309
hopyard, 57; dove dung for, 68, 74; weeding, 78; setting, 86; where and when to plant, 113; fee for tending, 286; looking after, 286

339

horses: strong and light, 31, 216; stabling, 48, 52; fodder, 57, 84, 242, 252; blood letting, 53, 57, 242, 318–19; feeding colts, 55; gelding, 70, 77, 248; plough horse, 80; lice, 239; gripes, 242; most suitable type of hay for, 273; bots, 298; mow-burnt corn, 304
hospitality, 58, 63, 65, 67
houses: thatched (reeded), 104; repairs, 127
housewife, attributes of good, 149
housewifery: note to reader on, 156; preface to book of, 157; praise of, 157; description of, 158; instructions on, 159–62; morning work, 163–72; afternoon work, 172–4; evening work, 174–5; supper time, 175; after-supper matters, 176–7; good and ill housewifery, 181–2
Huntingdon, 296
husbandry, good and ill, 132–3

implements, 31–3, 217–21; dull, 53; grindstone, 55, 240; whetstone, 55; fashioning, 56; protection of scythe (sieth), 76; hop tools, 86, 93, 264; gardening, 95; timber suitable for making, 98, 275; storehouse for, 112; for willow-propagation, 253; roller, 267; weeding, 284; borrowing of, 292
innkeepers, 184
Ipswich, xv, 205, 315
ivy, 74, 102, 104, 283

jackdaw (cadow), 95, 124, 271
justice, 136, 310, 311

Ket's Rebellion, 309, 311, 312
keys, 176
kitchen, seeds and herbs for, 88–9

labourers: cattle beaters, 50; nap, 108, 289; at harvest, 122, 124; housewife and, 160; roast meat for, 178; pigs of, 236; to be kept busy, 237, 291; day and piece labour, 302; feeding, 302–3; beer, 305; rewards after harvest, 305
Lancashire, cattle rearing, 247
land (*see also* common land): tillage of, 42–5, 103; open and enclosed, 43–4, 134–9, 231–2, 309, 310–11; breaking up, 70, 79; resting, 71, 230; stony, 99; sheep grazing, 105; not to be overgrazed, 107; manuring, 121, 230; ringed hogs in meadow and pasture, 224; neglected, 227; effect of football on, 243; sowing on unenclosed, 251; value of small enclosures, 257; drainage, 266; Tusser's attitude to enclosure of, 277; Lammas Lands, 309
latrines, cleansing, 48, 51
laundering, 169

learning, Hillman on, 288
leeks, 87, 95
Leicestershire, 135, 177, 178, 319
lending, 18, 19, 27
Lent: farmer's diet, 24, 318; provisions for, 85, 119; hungry dogs in, 92, 262–3; milch cow for, 119, 126; fish for, 125; not observed, 241, 306; Tusser and, 306–7
lentils, 80
lessons: husbandly, 16, 73; for a dairymaid, 100–1
Lincolnshire, 97, 247, 254
linen, 121, 299, 301
ling, 53
London: plague in, xvii, 209, 314; Tusser in, xvii, 208, 316; Tusser praises, 209; Billingsgate, 258; grass-cocking, 298
looking-glass, 280

magpies, 95, 110, 271, 291
mallow: destroying, 54, 68, 70, 74, 78; not to be dug in, 248
malt, 40, 46, 49
malting, 47, 49, 170
manure: and wheat crops, 44, 233; horse dung, 48; human excreta as, 51, 238–9; saving of, 52; compost, 55; mucking out, 68; taken to fields, 73; dove dung, 68, 74, 244; for barley, 80; dunghills to stand a month, 83, 250; plough after spreading, 83, 97; manure furrows, 83, 250; spreading, 103, 107, 307; after harvest, 124–5, 306; Hillman's recommendation, 228; various kinds of, 228; cost and profit, 228–9; for various kinds of land, 230; Middlesex, 232; yard to be cleaned of, 239; in frost, 244; cow dung and molecasts, 255; folding of land, 277; ploughing in, 287; where to make the dunghill, 287; mud from ditches, 296
marshes, 34, 56, 86, 92, 262
mast, 28, 36, 37, 224
meadows, 85, 97, 112
meat: for Easter, 24, 242; beef, 47, 50, 214, 308, 318; tainted, 178, 318; roast, 163; smoking and pickling, 237; high consumption of, 268
mending and making, 173
mice, 111, 168
Middlesex, 44, 232
milking: ewes, 76, 104, 247, 281; cow, 119; evening, 174
milling, 116, 300
mistletoe, 69, 74
molehills, 48, 51, 81, 85, 239
moles, 85, 239, 243, 255–7
money, 16
moon: phases of, 26; fruit gathering, 35, 223;

moon: *contd*
  bean and garlic setting, 50, 237; effect on tides, 56; sowing, 80, 83; gardening, 95; Prime, 247, 264; influence on agriculture, 251; grafting, 264

mowing: hay, 111, 291; meadows, 112; brakes, 121; barley, 123, 304

music: Tusser the chorister, xii–xiii, 203, 315; Lord Paget's musician, xiv; drums and horns at harvest time, 117; servants' singing, 166; teaching to children, 182

mustard: sowing, 80, 84; among hops, 106; reaping, 117; gathering and storing, 121, 302

myslen (mixed corn), 34

neighbours, 117, 140–1, 160, 251

nettles, 84, 248, 254

Norfolk: vetch threshing, 57; open country, 135; Tusser marries in, 206; light-fire, 235; turnip-fed cattle, 240; football, 243; cattle rearing, 243; twin lambs, 247; mustard seed, 254; barley land, 265; Red Raw, 273; reeding, 282; labourers' and dairymaids' rest, 289; ditch-mud, 296; haymaking, 297–8; work by moonshine, 305; brakes for fuel, 301; Sturbridge Fair, 308; enclosure of land, 309; woodland, 312; 'Norfolk wiles', 314

Northamptonshire, 178, 319

Northumberland, 304

Norwich, Tusser in, xvi, 207–8

nursery, 180

nutting, 35, 223

oak, 274, 308

oats: effect on land, 43, 79, 249; moist conditions, 45; sowing, 70, 78, 249, 267; harrowing, 94, 267; mixed with chaff, 217; fodder for horses, 252; to feed nesting swans, 255; wild, 266

orchards, 53, 56, 57, 75, 241–2

osier, *see* willow

oxen, *see* cattle

Paget, Lord William, Tusser's patron, xiii–xiv, 204, 313, 315; his portrait, xvi; death of, xvi; Author's Epistle to, 1–2; verses in praise of, 3–4

Paget, Lord Thomas, verses to, 3–6

Paget, Lady, 154–5, 189, 313

pasture: rooting of, 36; keeping clear, 75; ridding of turfs, 76; molehills on, 85; laying spare, 85; sheep, 104, 282; ringed hogs in, 224; ridding of ant hills, 246; management of, 307

peacock, 106, 253, 286

peas: sowing, 39, 45, 50, 80, 92, 102, 106, 234, 284; suitable soil for, 42, 229, 233; enrich soil, 43, 231, 233; in rotation of crops, 44; moist conditions, 45; hastings sown, 47; on fallow ground, 68; garden peas, 68, 253; for the pot, 73, 244; marrow-fat (runcivall), 74, 80, 83, 84, 253; times for sowing, 83, 84; harrowing, 83, 94, 251, 267; pottage, 85, 95; storage, 111, 118, 124; treatment, 123; weed-destructive, 226, 250; valuable commodity, 233; price, 236; new strains of, 244; influence of moon on, 251; profitability of, 262; affected by moisture, 293; turning, 304; covering, 305

pests: blowfly maggots (mads), 102; weevil (bowds), 170; nits, 52, 239; lice, 52, 239; fleas, 116, 300

physic, 114, 179, 313–14

pigeon (*see also* doves): scaring, 34; plague to farmers, 222; ravage peas and beans, 251; and hemp seed, 254; take peascods, 285

pigeon-house, cleaning, 74, 244

pigs: husbandry in September, 28, 35, 36; tools for pig husbandry, 33, 221; rooting, 36; ringing, 36, 71, 224; peasefed swine, 39; in October, 40; disease, 45; in November, 47, 48; pig meat, 47; reduce stock, 50; fattening, 54, 235; souse, 68; sows in farrow, 70; rearing, 73, 243; care of piglets, 70, 77; gelding, 70, 77; best farrowing time, 77, 247–8; to breed large hogs, 77; profit from 78, 248; swine to be yoked, 81; hogs and lambs, 86; shelter for, 111, 119, 125, 292, 306; hogs at cartwheel, 117, 122, 303; straying into woods, 223; hard and cool lying, 223–4; and propagation of thistles, 230; poor man's hog, 233, 277; slaughtering, 235; of day-labourers, 236; suckling, 248; growing sows, 248; damage caused by hogs, 277, 293, 310; hogs and the weather, 296, 306; finicky hogs, 293

planets, 26, 250–1

ploughing and fallowing: and sowing, 44; one or two crops to a fallow, 44, 233; November, 47, 48, 51; garden, 70, 77; delay, 71, 79; plan before, 80; dung spreading before, 83; stubble, 83, 250; May and April, 97; twifallowing, 103, 107; and mowing, 111; against weeds, 114, 115; thry fallowing, 115, 299, 301; summer, 215, 272; first ploughing of land for barley, 225–6; in drought, 287

ploughman: feasting days, 177–8, 319; hours of work, 289

ploughs: strong and light, 31; ploughing tools, 32, 219; crab-apple wood for, 96, 98; turn up stones, 246

poplar, 73, 75, 244
poultry, 47, 50, 88, 236, 304
poverty, 252

quagmire (quamier), 71

rabbit (conie), 68; cease killing, 74, 244; protection of burrows, 81, 250; browse, 245; to keep out, 310
rain, 41, 47, 111
raking, 118
rats, 168
ravens, 87, 95, 271
reapers, 122
reaping, 117, 118
Redford, John, xiii, 203, 315
reeding, 104, 282
reeds for thatch, 102, 104
religion: practical points of, 190; Tusser's belief, 191–5; omnipotence of God, 196
remedies (*see also* blood-letting): sloe, 40, 46, 235; verjuice, 40, 46, 69, 75, 235; bay salt, 75; soot and garlic, 75; dust, 104; basil, 108; wormwood, 116; tar, 220, 282, 290; for swollen throat in hogs, 234; for choking cattle, 240; scalded bran, 242; for disorders in horses, 245–6; herbs, 259–62, 268; scab water, 282
rolling, 87, 94, 265, 267
rooks, 251; scaring, 87; to be destroyed, 95; steal corn, 222; protected species, 271; take peascods, 285
roses, 27, 36, 68
rue, 114, 116, 261–2
rye: sowing, 27, 34, 39, 41, 267; straw, 40; suitable soil for, 42, 226, 229; impoverishes land, 43; in rotation of crops, 44; harrowing, 94; stubble, 250

saffron: preparing plot, 117; growing, 117, 121, 300; transplanting, 121, 300; linen bleaching on, 121, 300
salads, herbs and roots for, 89
Salisbury, John, Dean of Norwich, xvi, 208
sallow, *see* willow
salting bacon flitches, 47
sawpit, 29
'Scarborough warning', 18, 214
scouring, 169
scything (swinging), 117
seed: light, 31, 216; for kitchen, 88; exchanging, 117, 121, 302; saving, 121; picking and cleaning, 125; corn seed, 285; to prevent seeding, 299; mustard, 302; change of, 306
servants (creekes): bailiff, 56, 241; pottage for, 87; dairymaid, 100–1; master and, 115, 297;
housewife and, 160, 280; to breakfast, 164; mistress and maids, 165; chattering and brawling, 165; singing, 166; changing, 166; whispering, 169; at dinner, 170–1; waiting at table, 186; to be watched, 237, 278–9; to be kept employed, 239–40; punishment of, 279, 318; sluts, 280; Davus, 312; extravagance of, 317; standard of living, 318
shacktime, 36, 224
shearing, shears, 33; in June, 110; Northamptonshire, 178, 319; careless, 290; too early or too late, 290–1
sheep, 81; gelding, 27, 35, 70, 87, 247; tools for marking, 33; dogs and, 50, 86, 92, 263; mistletoe, 69; ewes in lamb, 69, 75; lambs, 69, 70, 75, 86, 87, 102; for meat, 73, 243; ewes' milk, 76, 104, 247, 281, 282; proportion of ewes to cows, 102, 104; mads, 102; folding, 104, 229, 277–8; laxative complaint, 104, 282; washing, 110, 290; shearing, 110, 178, 290–1; old ewes, 119, 127, 308; value of, 215; benefit from summer fallowing, 272; separation of lambs from ewes, 281; lambing time, 281; grazing, 282; sheepways, 310–11
shelter: cowhouse, sty and hogscote, 33; for cattle, 53, 55, 240; for ewes and lambs, 76, 246; for hogs, 111, 119, 125, 306; cart-shed, 111, 292; against storms, 265
shepherd, 96
sickness: plague, xvii, 209, 314; servants in, 172; physic, 179; Tusser ill, 204, 208; of Tusser's first wife, 205
slaughter: Hallontide, 49; boar, 54; rabbit, 74, 244; all year round, 235; at Prime, 247
sloes, 40, 46
soil (*see also* land): light, 31, 217; best kind of, 38; sand, 40; for various sorts of crops, 40, 229; indications of richness of, 42, 43, 230, 265, 284; peas enrich, 43, 231, 233; clay, 83, 239, 251, 296; for barley, 93, 265; for hops, 113, 248, 296; for rye and wheat, 226; exhaustion of, 231, 249–50, 266; for peas, 244; for poplar and willow, 244; effect of oats on, 249, 266; for propagating a quickset hedge, 252, 253; clods, 265, 266, 267; and putrefaction of weeds, 272; for grasses, 276; for flax, 299
soot, 238
Southwell, Sir Richard, xvi, 207, 315
sowing: September, 27, 29, 34, 221; October, 38, 41; and fallowing, 44; November, 50; January, 78, 79, 249; February, 80; March, 86–7, 92; kitchen garden, 94–5, 269; May, 102, 103, 105, 106; three converse ways of, 226–7; of headlands, 238; peas, 250; under

furrow, 251; on unenclosed land, 251; old rule on, 269; late and early, 270
spring, 25, 59, 245
stable implements, 31, 217–18
stones: gardens rid of, 78; thrown up by harrow, 88; gathering, 99, 103, 107, 278, 288; uses of, 99; thrown up by plough, 246
storehouse for tools, 112
storing: straw, 40, 51; grain, 48; peas, 111, 118, 124; mustard seed, 121; barley, 124, 305; wheat, 236; corn, 293; buckwheat, 300
straw (hawme, stover): storing, 40, 51; lay to rot, 48, 51; cattle fodder, 56, 80, 236, 241; saving, 81; burning, 117; getting in, 117; uses of, 122; reaping and mowing, 123; let lie in yard, 238; green peason, 242
stubble, 83, 250
Suffolk: Tusser in, xiv–xv, 205, 206, 228, 315; harrowing, 34; wheat growing, 42; enclosure of land, 135; Shrovetide, 177, 319; cattle rearing, 243; dairy farming, 278, 309; skimmed-milk cheese, 279; reeding, 282; Sturbridge Fair, 308
summer, 25, 59
sun: rising and setting of, 26; hop growing, 93, 106, 113, 263, 264; drying by, 169; south- and south-west-lying land, 268; hogs suffer in heat of, 292; and timber felling, 295
swans, 80, 85, 255
swine, see pigs
swineherd, 96, 99, 277

table lessons, 185
tanners, 98, 273–4
tares (tine), 250, 286
tenant farmers, 30
thatching (reeding), 104, 122, 282, 293, 303
thieves: dogs of, 29; of fruit, 35, 223; prowlers, 37; of corn, 47–8, 51; of eggs, 48; of fish, 82; of animals, 88; in Cambridge, 136–8; punishment, 225; at timber felling, 274; thieving from common land, 276; horse-stealers, 307
thistles, 105; factors encouraging, 42; among hops, 106; fallowing against, 114, 115; indicative of rich soil, 230, 265–6, 284; destruction of, 265, 284
thorn, 98, 112
threshers, 50, 51, 258, 283
threshing: September, 34; November, 47, 49, 50; February, 85; May, 105; inconveniences of, 283
thrift: the ladder to, 13–14; six noiances to, 18
tides, 26, 53, 56, 223, 241
tillage (tilthing), 42–6, 84, 87
timber: September work, 29, 37; lopping, 68; felling, 68, 75, 96, 98, 274, 295; saving, 69; sale of bark, 96, 98; stadling of woods, 98; use of various kinds of, 98; for fuel, 125; strength of, 216; shaken, 225; sawdust, 225; fence wood, 246
tithing, 21, 70; paying tithes, 78, 115, 248, 303; at harvest time, 117, 122; of barley, 123; antiquity of, 298
tomtits, 265
tools, see implements
Tottel, Richard, xi
trees (see also individual species): removing, 56, 72, 241; setting, 56, 71, 72, 241; pruning, 68, 74; lopping, 74–5, 245, 283; cutting back, 112; Dovercourt elms, 219; sap, 264; grafting, 264; pollards, 282–3
trees, fruit, 53, 71, 241–2
trees, nut, 71
trenching, 51, 93, 238
turkey, 106, 253, 286
turnips, 228, 236; and fertility of land, 229, 296; animal fodder, 240, 245; hoeing, 270
Tusser, Thomas: background, xi; education, xii–xiv, 203, 204, 315; Lord Paget's musician, xiv, 315; Suffolk farmer, xiv–xv, xvii, 205, 315; at West Dereham, xv–xvi, 207; under Southwell's patronage, xvi, 315; in Norwich, xvi, 207–8, 316; farms at Fairstead, xvi, 208, 315; takes refuge in Cambridge, xvii, 209, 316; dies, xvii, 315; Fuller on, xvii, xix, 317; first version of *Husbandrie*, xvii; its success, xviii–xix; its content, xix–xx; ill, 204, 208; at Court, 204–5; marriages, 205, 206; in London, 208–9, 316; religious belief, 191–5, 196, 215; his epitaph, 316; epigram on, 316; Scott on quality of his poetry, 320
Udall, Nicholas, xiii, xviii, 203, 315
usury, 20

vegetables: root (see also turnips), 238, 268, 269
verjuice (vergis), 40, 46, 69, 75, 245
vermin, 74, 285
vetch (fitch): to eke out hay, 57; threshing, 57; sowing, 83; in rye and wheat, 105; new strains of, 242; weed-destructive, 250; tine tare, 284; affected by moisture, 293
vines, 80, 85, 254–5

wages, 21, 237–8
walling, 34, 40, 48, 222
water: wheat-crops, 44, 45, 233; sheep washing, 110, 290; for setting plants, 269–70
water-furrough, 34, 42, 222, 227, 266
watering, 95, 107, 262
wattles, 248
wedlock, advice on, 15–16

343

weeding: in May, 102, 103, 105, 106; oak and hawthorn seedlings, 233; of quickset, 244, 286–7; new hedges, 254; thistles, 265; wild oats, 266; implements for, 283–4; of corn, 284; of wheat, 286

weeds: among barley, 93, 265–6; in a garden, 94; titters and tine, 102, 103; among hops, 106; fallowing against, 115, 272, 299, 301; destroyers of, 226, 250; a list of, 284; danger of neglecting, 285

wheat: sowing, 34, 38, 39, 41, 102, 226; straw, 40; in Suffolk, 42; types and quality of, 43, 231; impoverishes land, 43; growing conditions, 44–5; old better than new, 46; threshing, 49; rolling, 94, 267; weeding, 106, 286; seed, 118; a coome, 218; white and red, 222; Pride and Poverty, 230; sweating in mow, 235; to lie before threshing, 236; to preserve in granaries, 236; stubble, 250

willow: planting, 54; setting, 68; for hedging, 73; sallow, 75, 275; cutting osier, 80, 85; to cast shade, 84; hops and, 86; for hop poles, 127; best type of soil for, 244; to propagate, 253; growing conditions, 255; for fencing, 263; use of timber, 275

Wiltshire, 253

winds, 25, 113, 169, 264–5, 306

winnowing, 47, 118

winter, 25, 59

wood, *see* timber

woodland, 30, 216, 225, 311

woods, 45, 98, 233–4, 274

wool, 27, 35, 290, 291

wormwood, 114, 116, 300